Biochemical Mechanisms
of Detoxification in Higher Plants

George Kvesitadze · Gia Khatisashvili
Tinatin Sadunishvili · Jeremy J. Ramsden

Biochemical Mechanisms of Detoxification in Higher Plants

Basis of Phytoremediation

With 96 Figures and 8 Tables

Springer

George Kvesitadze
Gia Khatisashvili
Tinatin Sadunishvili
Durmishidze Institute of
Biochemistry and Biotechnology
David Agmasheneblis Kheivani, 10 km
0159 Tbilisi, Georgia
gkvesitadze@yahoo.com

Jeremy J. Ramsden
Chair of Nanotechnology
School of Industrial and Manufacturing Science
Cranfield University
Bedfordshire
MK43 0AL, UK
j.ramsden@cranfield.ac.uk

ISBN-10 3-540-28996-8 Springer Berlin Heidelberg New York
ISBN-13 978-3-540-28996-8 Springer Berlin Heidelberg New York
e-ISBN 3-540-28997-6

Library of Congress Control Number: 2005932549

Springer is a part of Springer Science+Business Media
springer.com

© Springer-Verlag Berlin Heidelberg 2006
Printed in Germany

Typesetting and Production: LE-TEX, Jelonek, Schmidt & Völcker GbR, Leipzig
Coverdesign: design&production, Heidelberg

Printed on acid-free paper   2/YL – 5 4 3 2 1 0

Preface

In our new millennium environmental problems have become closely linked to everybody's life: the condition of the environment has become one of the most vitally important parameters determining the continuing existence of mankind. The inexorable and seemingly permanent increase of technogenic contaminants in all ecological niches creates danger for all nature. Confronted with this situation, the rational use of the capabilities of higher plants, which even today, after the anthropogenic destruction of so much flora, still cover more than 45% of the land, to absorb and detoxify contaminants of amazingly diverse chemical structures might be a significant route to solve the problem. As an inherently biological principle of environmental remediation, phytoremediation is the closest to nature among the various available technologies, and is especially effective against widespread contamination of low concentration (chronic pollution), which, it is now recognized, is a far more serious challenge to the overall world environment than the dramatic accidents that the media are so fond of highlighting.

This book has been written to provide an overview of the fundamental aspects of phytoremediation, to summarize existing understanding of the mechanisms of detoxification of environmental contaminants in plants; to describe the principles of the practical realization of these modern technologies; to show the degree of disturbance to plant cell homeostasis under the action of toxicants at different doses; and to rationally evaluate the ecological potential of plants.

There is no doubt that all kinds of ecotechnologies based on mechanical, chemical, physical and biological principles of environmental preservation are important at different levels, such as within an individual company or chemical plant, a district, country, or the entire globe. Worldwide environmental defense calls for the partnership of all countries, since pollution rarely respects political boundaries, and purely national programme are unlikely to be very effective. The creation of international projects in environmental preservation should lead to closer international cooperation on issues extending beyond the physical and biological environment, which is expected to have indirect benefits in many fields.

The aim of the authors is to convey the framework upon which the comparatively new multidirectional discipline of phytoremediation is based. The new and selective use of vegetation for the decontamination of polluted sites from organic and inorganic contaminants includes several aspects of plant physiology, plant biochemistry, organic and inorganic chemistry, microbiology, molecular biology, agronomy, engineering, etc.

The essence of phytoremediation lies in understanding the pathways along which environmental contaminants enter plant cells, and then how those cells deal with the xenobiotics. The preferred fate for organic contaminants is to be transformed into molecules that can enter the regular metabolic channels of the plant; in other words, the proper metabolism of organic contaminants actually could provide nutrition to the plant. Depending on the conditions, however, especially the dose of the contaminant, it may merely be conjugated with a suitable endogenous compound available within the cell and temporarily passively stored.

In order to understand the mechanisms of degradative transformation in higher plants, only a small number of enzymes need to be considered, chief among which are the cytochrome P450-containing monooxygenases, peroxidases and phenoloxidases. It is very important to realize that a plant cell is not a miniature factory, in the way that some bacteria are. An individual cell has a limited capacity to eliminate a xenobiotic contaminant. Xenobiotic-induced changes in gene expression and cell ultrastructure are discussed in this book. Despite this limitation, phytoremediation technologies can be successfully applied since plants can rapidly grow and multiply to occupy large volumes.

This book has a distinctively practical aim and is intended to serve as a working handbook for anyone involved in setting up phytoremediation defenses in order to improve environmental quality and hence the quality of life for human beings. The authors' credo, however, is that this work will be far more effective with a deeper appreciation of the molecular mechanisms within the plant cell underlying the process and its overall limitations.

The authors express their sincere thanks to Prof. Friedhelm Korte (Munich Technical University, Germany) for his long-term scientific collaboration from which this book has profited to a great extent. It is also their pleasure to express thanks and appreciation to Dr Elly Best (U.S. Army Engineer Research and Development Center, Vicksburg, Mississippi, USA) for her critical reviewing of the individual chapters and for making useful suggestions. Authors express their thanks to Mrs Enza Giaracuni for having made sense of our highly convoluted and overwritten drafts and produced a pristine final typescript.

Tbilisi, August 2005   Giorge Kvesitadze

Table of contents

Acronyms

ADNT	Aminodinitrotoluene
ATSDR	Agency for Toxic Substances and Disease Registry
BTEX	Benzene, toluene, ethylbenzene, xylenes
2,4-D	2,4-Dichlorophenoxyacetic acid
DCA	3,4-Dichloroaniline
DDT	Dichlorodiphenyltrichloroethane
DNOC	Dinitro-*o*-cresole
EPA	United States Environmental Protection Agency
FDA	United States Food and Drug Administration
GSH	Glutathione
GST	Glutathione S-transferase
HMX	Octahydro-1,3,5,7-tetranitro-1,3,5,7-tetrazocine
IARC	International Agency for Research on Cancer
IPSC	International Programme on Chemical Safety
MTBE	Methyl *tert*-iarybutyl ether
PAH	Polycyclic aromatic hydrocarbons
PCB	Polychlorinated biphenyl
PHC	Petroleum hydrocarbon
POP	Persistent organic pollutant
RAM	Risk assessment and minimization
RDX	Hexahydro-1,3,5-trinitro-1,3,5-triazine
2,4,5-T	2,4,5-Trichlorophenoxyacetic acid
TAT	Triaminotoluene
TCDD	2,3,7,8-Tetrachlorodioxybenzodioxin
TCE	Trichloroethylene
TNT	2,4,6-Trinitrotoluene
TSCF	Transpiration stream concentration factor
VOC	Volatile organic compound
WHO	World Health Organization

1 Contaminants in the environment

1.1 Environmental contaminants

The constant increase of environmental contamination by chemical compounds is one of the most important and unsolved problems burdening mankind – a XXI century sword of Damocles. The sources of chemical pollution are divided into two types, natural and anthropogenic, i.e., originating from human activity. Natural contamination may result from elemental processes such as the emission of poisonous gases during a volcanic eruption, the washing of toxic elements out of ore during floods or earthquakes, as well as the metabolic activity of all kinds of organisms, excreting toxic compounds, etc. In comparison with nature, however, the human contribution to environmental contamination is much more impressive. As a result of urbanization, the unpredictable growth of industry and transport, the annual increase of chemical production, military activities, etc., man has created a multi-barrelled weapon that finally appears to be targeted back onto him.

More than 500 million tons of chemicals are produced annually in the world. In different ways, huge amounts of these hazardous substances or toxic intermediate products of their incomplete transformations are accumulated in the biosphere, significantly affecting the ecological balance. Nevertheless, members of the plant kingdom (lower microorganisms and higher plants) can assimilate environmental contaminants, and be successfully directed to remove toxic compounds from the environment, providing long-term protection against their environmental dispersal in ever increasing doses [278]. Lately, many ecological technologies have been elaborated, targeted to minimize the flow of toxic compounds to the biosphere or to control their level or state [505]. Despite the definite positive effect from the realization of these technologies, the intensive flow of toxic compounds to the biosphere is still increasing.

Toxicity (from the Greek *toxikon* – poison for oiling arrows) is the ability of natural and synthetic chemical compounds, in doses exceeding normal pharmacological levels, to induce disruption of normal vital processes. In cybernetic terms, one would say that the organism's essential variables are pushed towards their limits, and sometimes beyond them. Toxicity is revealed in different ways. At the contemporary stage of development of medicine, biology and chemistry toxicity does not have a precise definition. Operationally, the term toxicity can be defined as the action of compounds characterized by the ability to suppress or significantly inhibit the growth, reproduction or definite functions of some organisms, or particular organs within them [287].

Undoubtedly, the great majority of chemically synthesized compounds such as plant protection and pest control agents (pesticides), paints and varnishes, solvents and emulsifiers, petroleum products, products of applied chemistry, many chemicals widely used in the polymer industry (monomers, dyes, plasticizers, stabilizers, etc.), products of the pharmaceutical industry, surfactants, refrigerants, aerosols, explosives, heat-generating elements of nuclear power stations, conservation agents and packing materials, are toxic. Some of these products are characterized by a much higher toxicity than otherwise comparable natural compounds. Examples of exceptions (highly toxic compounds of natural origin) are cyanogenic glycosides, glucosinolates, glycoalkaloids, lectins, phenols, coumarins and some other secondary metabolites of plants. Toxins of microorganisms are specific poisons elaborated by both prokaryotic and eukaryotic microorganisms. These toxins are often polypeptides varying in molecular mass and may contain up to one hundred thousand amino acids [118]. Low-molecular-mass organic compounds are also encountered as microbial toxins. In spite of their high toxicity, these compounds exist in nature at such low concentrations in comparison with anthropogenic toxic compounds that they cannot be considered as contaminants.

Waste often contains contaminants of low toxicity. Industrial wastes are substances (chemical compounds) that cannot be processed within the framework of existing technology, or their further treatment is economically inefficient [287]. The main point of any technological process is an application of different types of action such as physical (e.g., mechanical), chemical, biological, or combinations thereof. Often, after exhausting all possible effects of one type of treatment, the use of other types may allow further transformation. Due to the intensification of industry, driven by an exponential growth of population, wastes have become extremely numerous. Wastes often serve as a source for the development of toxic microflora, which can transform them into hazardous contaminants. The wastes of agriculture and the food industry can be classified as substances of low toxicity. Even some components of food (alcohol, fats, aldehydes, secondary metabolites, other ingredients) might be considered as compounds having a (very low) toxicity, although their consumption in small amounts is harmless. The United States Food and Drug Administration (FDA) has created and maintains a list of common plant toxins, and guidelines defining acceptable toxin levels. A number of medical supplies, including antibiotics, also belong to the category of substances of low toxicity.

In the remainder of this chapter, several classes of highly toxic environmental contaminants and some individual compounds will be discussed in more detail.

1.1.1 Pesticides

Pesticides, compounds for plant protection and pest control, are ranked as the most widely distributed chemical contaminants of the environment in the twentieth century. According to data compiled by the United States Environmental Protection Agency (EPA) and the World Health Organization (WHO), over 1000 compounds are used as pesticides, representing compounds of many different chemical classes: carbamates, thiocarbamates, dipyridyls, triazines, phenoxyacetates, coumarins, nitrophenols, pyrazoles, pyrethroids, and organic compounds containing chlorine, phosphorus, tin, mercury, arsenic, copper, etc. Millions of tons of pesticides are produced and used annually in close association with agriculture. Many articles, reviews and books are devoted to pesticides [375, 376]; here only the sources of pesticides, their distribution and toxicity will be reviewed.

Pesticides are generally divided into the following groups according to their type of action:
- Acaricides (also called miticides), agents used against mites.
- Algicides, to combat algae, applied for the sanitary control of lakes, channels, water pools, water reservoirs, etc.
- Antifouling agents, against different organisms growing on the underwater surfaces of boats and ships.
- Attractants, the means to entice parasites, insects and rodents into special traps.
- Bactericides or biocides, disinfectants and sanitizers for the annihilation of some pathogenic microorganisms.
- Defoliants, the means for rapid defoliation, generally used to facilitate the harvest of useful crops.
- Desiccants, chemicals promoting desiccation of tissues, e.g., the roots of undesirable plants.

- Fumigants, the means to produce gas or vapor to destroy pests in buildings or in soil.
- Fungicides, agents against blights, mildews, mold and rusts.
- Herbicides, to combat weeds and toxic vegetation.
- Insecticides, chemicals for the annihilation of harmful insects and other arthropods.
- Molluscicides, agents for protecting submarine surfaces from snails and slugs.
- Nematocides, the means to protect from nematodes, worm-like microscopic organisms that feed on plant roots.
- Ovicides, chemicals to damage the eggs of insects and mites.
- Pheromones, agents against the propagation of insect populations.
- Plant growth regulators, to deliberately change the growth rate, flowering and reproducibility of plants.
- Repellents, agents to repel pests, including insects (especially mosquitoes) and birds.
- Rodenticides, the means for protection from rodents.

Due to the widespread and long-term application of pesticides in agriculture, soils, ground waters and reservoirs in many areas are now heavily contaminated. The toxicity of pesticides makes them hazardous when incorporated into the food chain.

Examples of the most widely distributed pesticides are given below:

Alkylphosphates   Parathion

Barban – 4-chloro-2-butynyl (3-chlorophenyl)carbamate

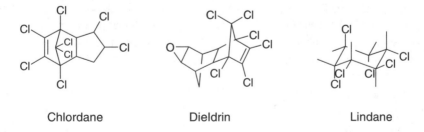

Betanal – 3-((methoxycarbonyl)amino)phenyl(3-methylphenyl)carbamate

$$\left[-S-\overset{\overset{\displaystyle S}{\|}}{C}-NH-CH_2-CH_2-NH-\overset{\overset{\displaystyle S}{\|}}{C}-S-Mn- \right]_n$$

Maneb – ((1,2-ethanediyl-bis(carbamodithioato))(2-))manganese

Alkylphosphates (triethylphosphate) are strong inhibitors of acetylcholinesterase. This affects the transmission of signals to nerve endings via the acetylcholine receptor. The decrease of enzymatic activity leads to an accumulation of symptoms of diseases such as sialorrhea, pulmonary edema, colics, diarrhoea, nausea, weakening of sight, rise of blood-pressure, muscular spasms and convulsions, speech disturbance, and paralysis of the respiratory tract. A similar clinical picture is observed in the case of phosphates and carbamates administered at intentionally increased doses.

Organochlorine insecticides (e.g., chlordane, dieldrin, lindane, DDT) typically penetrate into the human organism through the digestive tract or the skin [207]. When the membranes of nerve cells are damaged by pesticide action, their permeability for osmotic transport of Na^+-flow is maintained. Hence, their rest potential after excitation either does not return to its initial level, or is decreased. These organochlorine compounds severely change the excitability of nerve cells. At low concentrations axons are damaged; higher concentrations also cause the damage of sensory neurons. Chlordane and dieldrin are, moreover, clearly carcinogenic.

Chlordane   Dieldrin   Lindane

Dichlorodiphenyltrichloroethane (DDT) is an extremely active insecticide. This compound was first synthesized in 1874, and since 1930, when its insecticidal properties were established, was widely used against the malaria transmitter, the *Anopheles* mosquito [276].

Dichlorodiphenyltrichloroethane (DDT)

The practically unlimited application of DDT led to its worldwide distribution. Its high solubility in fat favored its incorporation into food chains. As a result, in the terminal steps of food chains the concentration of DDT is typically increased almost a million times, e.g., starting from rainwater and ending in human milk [162].

DDT is well absorbed by clays, is accumulated in humus rich in pine needles, where it is dissolved in the wax of the needles. Destroying many organisms, this compound negatively influences ecosystems. It is a typical contact poison, rapidly penetrating through the skin. DDT induces apoptosis in human mononuclear cells *in vitro* and is associated with increased apoptosis in exposed children [373].

It induces DNA damage in blood cells [539], and adversely affects the normal duty cycle of nerve cell membranes, as it depresses the response of the Na^+-pump, hence normal restoration of the resting potential does not occur after excitation of the nerve; large amounts of DDT induce paralysis of extremities. Maternal milk containing the insecticide can seriously damage the health of a child or disturb latent reproductive capacity by penetrating into the gonads.

Under usual conditions, DDT slowly and partially decomposes. Under aerobic conditions decomposition products are derivatives of dichloroethylene, which are less toxic than DDT; under aerobic conditions the dichloroethane derivatives are formed, which are easily transformed into derivatives of acetic acid [162].

Chlorinated herbicides have specific physiological effects on humans. For example, 2,4-dichlorophenoxyacetic acid (2,4-D) and 2,4,5-trichlorophenoxyacetic acid (2,4,5-T) act typical for herbicides to a less extent than their admixture with 2,3,7,8-tetrachlorodioxybenzo-*p*-dioxin (TCDD). The toxicity of the latter compound is 500,000 times higher than that of 2,4-D.

Even if the TCDD content in a herbicide is only 0.005 mg/kg, this concentration cannot be considered as harmless, because TCDD exerts not only very high toxicity under natural conditions, but is also extremely stable.

2,4-D   2,4,5-T   2,3,7,8-TCDD

In 1971, in the small town called Times Beach, Missouri about 10 m^3 of technical oil was splashed onto the ground of a hippodrome to avoid dust at horse races. Several days later the hippodrome was covered with dead birds, and a day later three horses and a rider fell ill. 29 Horses, 11 cats and 4 dogs died within one month. After 3 months several adults and children fell ill [90]. The authorities were forced to investigate and found dioxins and furans at concentrations of 30–53 ppm in the ground. The technical oil used at the hippodrome was waste from the industrial production of 2,4,5-trichlorophenol and contained TCDD. 2,4,5-Trichlorophenol is used as an insecticide and also is a reaction component in the production of 2,4,5-T ("agent orange"). 2,4,5-Trichlorophenol and 2,4,5-T can be easily transformed into TCDD by elimination of two molecules of hydrogen chloride and two molecules of chloroacetic acid.

Dipyridyls, such as the herbicide paraquat, induce blisters and ulcers even upon slight external contact with the skin [6].

Paraquat

Paraquat may act synergistically together with iron to provoke Parkinson's disease [6]. After penetration into an animal, dipyridyls damage kidneys and liver and then cause fibrous changes in the lungs, often with lethal consequences. Due to their high toxicity, dipyridyls require especially careful handling.

The so-called pyrethroid pesticides also have toxic characteristics and are synthetic analogues of a widely distributed insecticide pyrethrin, a natural compound found in chrysanthemums. Pyrethrins intended for insecticide use are modified to increase their stability. Synthetic pyrethrins have toxic effects on the nervous system [475].

Pyrethrin

1.1.2 Dioxins

The group of polychlorinated dibenzodioxins and dibenzofurans, called dioxins is distinguished by an especially high toxicity [90]. These compounds are always found as a complex mixture.

Polychlorinated dibenzo-*p*-dioxin   Polychlorinated dibenzo-*p*-furan
n is the number of chlorine atoms and varies
from 4 to 8 for the entire molecule

The basic sources of dioxins are chemical factories producing chlororganic pesticides, polychlorinated chlorobenzenes, solvents for a number of chlorine-substituted alkanes (mainly dichloroethane, trichloroethane, ethylene chlorohydrin), and chlorine substituted polymers (above all polyvinyl chloride). Dioxins are also found in the gas used in the chlorination of water supplies. Polychlorinated contaminants are formed as admixtures during the interaction of chlorine with carbon (e.g., of electrodes and aerial oxygen). It should be stressed that dioxins constitute a serious hazard to the environment and human health [90].

According to the rate of environmental contamination by dioxins the pulp and paper industry is the second worst after the chemical industry. For the production of paper from woody raw material, lignin must be eliminated to leave just the cellulose. This delignification process implies the interaction of phenolic lignin fragments with chlorine reagents resulting in the formation of dioxins or their precursors. During the standard bleaching process of woody polysaccharides, chlorine or its derivatives are used, leading to the formation of polychlorinated contaminants.

Dioxins are also formed in high-temperature chemical processes (including garbage incineration) in which organic and inorganic compounds with one or more atoms of chlorine (including molecular chlorine) participate. Automobiles are another source of dioxins. These compounds end up in the exhausts along with the combustion gases from the engines of cars working on fuel containing tetraethyl lead (antiknock agent) and 1,2-dichloroethane (added to reduce lead accumulation inside the engine). Like other polychlorinated compounds, dioxins do not undergo transformation after entry into humans. Their penetration often leads to the development of chloracne, a severe skin disease, followed by long-lasting open sores and damage of the endocrine system, adversely affecting proper sexual development and usually fatally affecting embryos. Dioxins cause immunodeficiency, increasing sensitivity to infectious diseases, and are of carcinogenic nature. Comparison of the minimal lethal doses with the semi-lethal doses of dioxins shows their high toxicity. The toxicity of TCDD is 3.1×10^{-9} mol/kg, while for the toxin curare it is 7.2×10^{-7}, for strychnine 1.5×10^{-6}, for sodium cyanide 3.1×10^{-7}, for diisopropyl fluorophosphate (a chemical warfare agent) 1.6×10^{-5} mol/kg. Only lethal doses of the toxins formed by the pathogenic botulism bacteria (3.3×10^{-17}) and diphtheria (4.2×10^{-12}) exceed the toxicity of dioxins [171].

In spite of their high resistance, dioxins do appear to undergo very slow biodegradation. In the literature there are no data reporting the ability of plants to transform dioxins, but some microorganisms are able to mineralize these harmful toxic compounds. The eventual degradation of dioxin molecules is conducted by the joint action of aerobic and anaerobic microorganisms. Anaerobes, in particular *Dehalococcoides* sp. strain CBDB1, carry out reductive dehalogenation of dioxins [59] leading to the formation of *p*-dioxins. The latter compounds undergo enzymatic transformation by the aerobes *Sphingomonas* sp. RWI with participation of dioxygenases and hydrolases, as a result of which the splitting of the aromatic ring is observed (Fig. 1.1) [147]. Some microorganisms, for example, soil microscopic fungi and actinomycetes, are sensitive to the effect of dioxins, and the absence of these taxonomic groups of microorganisms in soil can serve as a bioindicator of the level of contamination by dioxins [343].

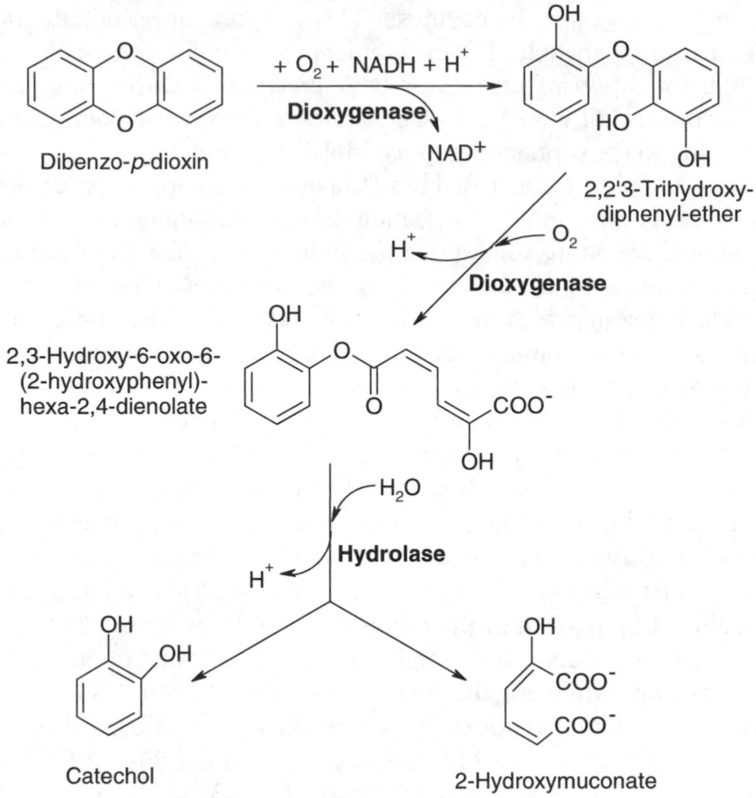

Fig. 1.1. Microbial degradation of dibenzo-*p*-dioxin

1.1.3 Polychlorinated biphenyls (PCBs)

PCBs, a family of over two hundred compounds, are characterized by an extremely high toxicity among polychlorinated aromatic compounds. Due to their high flame resistance, PCBs are used as anti-flash additives in electronics, printing equipment, transformer capacitors, packing materials, and as plasticizers in plastics, and as liquid thermoformer components of many technical oils. PCBs are antidusting agents of pesticides.

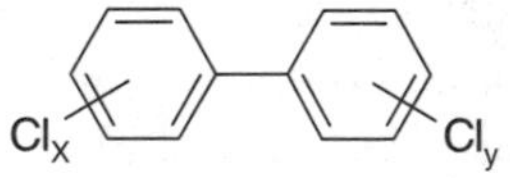

PCBs. Aroclor–1254 has x+y=5, for example

Though PCBs are slightly soluble in water and have a high boiling point, they are widely distributed in air, water and soil. The present general environmental contamination with PCBs is connected with their wide application. Despite strict restriction of the application of PCBs in industry, they are found in large amounts in soils and sediments, and via this pathway in water and air. Due to their high chemical stability and lipophilicity PCBs tend to be stable under natural conditions and remain unchanged for a long time. If a PCB molecule contains chlorine not exceeding 30% of its total mass it can be more easily removed by organisms than PCBs with a higher halogen content (60% or more) [162]. PCBs accumulate in plants and animal tissue, are inserted into the food chain, and therefore are ultimately dangerous for human health [206].

PCB toxicity increases proportionally with the increase in chlorine content. Poisoning by PCBs causes chloracne, changes the blood composition and affects the liver and the nervous system. There is considerable evidence for the carcinogenic nature of these compounds.

PCB residues are hard to annihilate. The best method is burning at a temperature above 1200 °C. They belong to that category of substrates whose application must undoubtedly be restricted. According to the literature, only a few bacterial strains are able to perform the full mineralization of PCBs as a result of aerobic and anaerobic conversions [269]. This process is slow compared with the rate of microbial or enzymatic degradation of many natural and even synthetic compounds. Initially dehalogenation of PCBs molecule takes place and the aromatic biphenyl rings become accessible for the oxidizing enzymes that degrade toxic compounds to regular cell metabolites. Data on the ability of plants to degrade PCBs are very exiguous [76].

1.1.4 Polycyclic aromatic hydrocarbons (PAHs)

As PCBs, the aromatic hydrocarbons contain condensed rings. PAHs are almost insoluble in water, have high boiling points and are difficult to decompose [162].

3,4-Benzopyrene   1,2-5,6-Dibenzanthracene

7,12-Dimethyl-
benzanthracene   3-Methylcholanthrene   3,4-Benzofluoranthrene

All these compounds have at least one cavity (marked by arrows) in their molecular structure. This feature is a characteristic for many carcinogenic compounds. No reliable information is available on PAH release on an industrial scale. Compounds of this class are formed during combustion of all kinds of fuel, and many natural products contain them. PAHs can be found in pitches, bitumen, soot, and humus components of soil. They are components of the exhaust gases of engines, combustion products of cooking stoves or space heating furnaces, smoked foods, tobacco and a number of other natural and anthropogenic products. PAHs are widely distributed and very stable under practically any conditions, creating a real danger to accumulate in living organisms at elevated concentrations.

Carcinogenicity of PAHs in mice has been reported [98]. After penetration of PAHs into the organism enzymes form epoxy compounds. These compounds react with guanine and block DNA synthesis, inducing disablement of transcription processes, or leading to mutations, which often promote cancer. There is substantial data in the literature indicating the potential of microorganisms and plants to degrade PAHs to regular cell metabolites [139, 450, 502, 504].

1.1.5 Phthalates

Phthalates, esters of phthalic acid, form another group of aromatic ring-containing toxic compounds. These esters are used as softeners in the production of polyvinyl chloride and other polymeric materials. Esters of phthalic acid are widely used in the production of solvents, lubricants, pesticides, lacquers and dyes, paper, perfumery, etc. [162].

Ester of phthalic acid

In plastics, these esters may comprise up to 40% of the mass. Phthalates are found in soil, air, and water. The primary source of phthalate distribution is the loss during their production processes. Besides, in time they slowly diffuse from plastics. Esters of phthalic acid are characterized by slight solubility in water and insignificant volatility. Upon burning, they volatilize. Phthalates are adsorbed onto organic materials in soils, leading to their accumulation in water reservoirs mainly on sediments. They also accumulate in sewage. Phthalates can be found in food products with artificial wrapping in very low concentrations (ppm).

Only an insignificant proportion of the ingested phthalates is absorbed via the gastrointestinal tract. They affect the skin and mucous membranes, slightly irritating them. Toxic effects of these compounds on organisms have not been studied comprehensively, although there are indications that the most widely distributed (80% of all phthalates used) compound, dioctyl phthalate (*bis*(2-ethylhexyl) phthalate), is carcinogenic [160]. On the other hand, the harmful action of phthalates on plants has been documented: it causes chlorosis (fading of the green coloring of leaves). Obviously, in this particular case the action of phthalates is connected with the disruption of chlorophyll biosynthesis [162].

Phthalates undergo degradation by microorganisms and plant enzymes [259]. As a result of bacterial action, the phthalic acid is formed initially from phthalates that are decarboxylated by ring splitting after hydroxylation. At the end of this process succinate and CO_2, or pyruvate and CO_2, completing the natural decomposition of glucose, are formed. The biological degradation of a phthalate molecule lasts several weeks on average.

1.1.6 Surfactants

Surfactants or detergents (tensides) create huge problems of water pollution. They are used as washing enhancers, reducing water surface tension. Their use is often followed by foaming [162]. Surfactants are organic compounds with both hydrophilic and hydrophobic moieties, and they fall into several distinct classes. The most widely environmentally distributed surfactants are the alkylsulfonic acids, the sulfuric acid residue of which forms the hydrophilic moiety:

$$R-\underset{\underset{OSO_3^-}{|}}{C}-R$$

In the case of the polyoxyethylenes, compounds of nonionic character, alcohol groups form the hydrophilic part of the molecule. Polyoxyethylene can form an ester with the residue of a fatty acid, or an ether with the residue of a high-molecular-mass alcohol:

$$RCOO- (CH_2 - CH_2O)_nH \qquad\qquad RCH_2O- (CH_2 - CH_2O)_nH$$

$$\text{Ester} \qquad\qquad\qquad\qquad \text{Ether}$$

Alkyl ammonium compounds contain positively charged quaternary ammonium as the polar (i.e., hydrophilic) component. Therefore they are called inversion soaps. Their bactericidal action is noteworthy.

$$R_1 - N^+(CH_3)_2 - R_2$$

Increasing industrial demands for surfactants and their intensive application in everyday life has led to widespread accumulation of foam in rivers and reservoirs. Foam hampers navigation, and the toxicity of surfactants causes mass extermination of fish. Negative experience in surfactant exploitation in the 1950s forced the search for biodegradable surfactants. Those having an unbranched chain, as e.g. nonionic detergents with alkylbenzene sulfonates, have this property [548]:

$$[CH_3(CH_2)_n -C_6H_4 -SO_3^-]R^+$$

These compounds are furthermore characterized by low toxicities for humans and fish [499]. Biotic disintegration of the chains in their molecules is accomplished by β-oxidation, i.e., by splitting acetic acid residues. Even very low concentrations (0.05–0.1 mg/l) of surfactants in rivers can become toxic, however: surfactants adsorbed to sediments and waste-water containing tensides can lead to the activation of toxic substances hazardous for groundwater. It is clear that the search for tensides of biogenic origin that can be rapidly and totally biodegraded must be continued and is of prime importance.

1.1.7 2,4,6-Trinitrotoluene (TNT)

A very toxic contaminant, TNT, is used as an explosive compound and intermediate in the production of dyes and photographic materials.

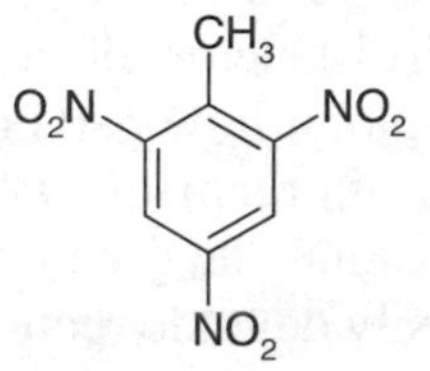

$$CH_3$$
$$O_2N \qquad NO_2$$
$$NO_2$$

2,4,6-Trinitrotoluene

The production and use of TNT for military purposes has led to its wide distribution. TNT is one of the most toxic explosives in the military arsenal, and has contaminated thousands of hectares of land. TNT mobility in soil is limited due to its strong adsorption onto soil particles.

TNT, assimilated via the digestive tract, skin and lungs, is accumulated mainly in the liver, kidneys, lungs and adipose tissue, stimulating chronic diseases [365]. According to EPA data, TNT is classified as a carcinogenic substance of group C. In animals TNT is slowly metabolized via the reduction of nitro groups leading to the formation of nitroso derivatives, hydroxylaminodinitrotoluenes, aminodinitrotoluenes (ADNTs) and diaminonitrotoluenes. Besides, oxidation of the methyl group and formation of nitro- and amino-derivatives of benzyl alcohol and benzoic acid may take place. Some of these metabolites (mainly amino-derivatives) are conjugated with glucuronic acid [146]. Formation of nitroso and hydroxylamino groups is the factor predetermining the toxic effect of TNT on an organism [21]. These groups bind to cell biopolymers, including nucleic acids and finally lead to chemical mutagenesis [408].

Microbial transformation of TNT usually begins with reduction of one of the nitro groups. The enzymes that catalyse these reductions are nonspecific NAD(P)H-dependent nitroreductases [156]. Complete reduction of nitro groups significantly reduces the mutagenic potential and toxicity of TNT.

Microorganisms degrade TNT in the following ways:
- Elimination of nitrogen in the form of nitrite and further reduction of nitrite to ammonium under aerobic conditions.
- Reduction of nitro groups by nitroreductase under anaerobic conditions and further aerobic metabolization of amino derivatives.

There are data indicating that TNT can serve as a terminal electron acceptor in the respiratory chain and that TNT reduction is coupled with ATP synthesis [156].

Some strains of *Pseudomonas* and representatives of mycelial fungi are able to utilize TNT as a source of nitrogen and carbon, and incorporate atoms of these elements in the skeleton of regular cell metabolism compounds. This is a good example of how parts of toxic compounds can participate in the vital processes of organisms. *Phanerochaete chrysosporium* and some other strains of basidial fungi completely mineralize TNT. The enzymes of basidial fungi easily degrade reduced metabolites of TNT. Due to the high intra- and extracellular activities of woody polysaccharide-degrading enzymes (cellulases, hemicellulases), and such oxidases as lignin peroxidase, Mn-peroxidase, and laccasey the strains of *Phanerochaete chrysosporium* are characterized by a high degradational potential.

The ability to absorb and assimilate TNT is intrinsic to different kinds of plants. The aquatic plant parrot feather (*Myriophyllum aquaticum*) and the alga stonewort (*Nitella* sp.) are used for the remediation of TNT-contaminated soil and water. Enzyme nitroreductase, which reduces TNT nitro groups, is active in other algae too, in ferns, and in several monocot and dicot plants, e.g., poplar (*Populus* sp.) trees [151]. Transgenic tobacco (*Nicotiana tabacum*) with the expressed gene of bacterial nitroreductase acquires the ability to absorb and eliminate TNT from the soil of military proving grounds [213].

1.1.8 Chlorinated alkanes and alkenes

Chlorine-substituted alkanes and alkenes (e.g., tetrachloromethane (CCl_4), dichloromethane (CH_2Cl_2), chloroform ($CHCl_3$), dichloroethane (CH_2Cl-CH_2Cl), vinyl chloride ($CH_2=CHCl$), trichloroethylene ($CCl_2=CHCl$), tetrachloroethylene ($CCl_2=CCl_2$), etc. deserve much attention among the toxic derivatives of hydrocarbons. These substances are often used as solvents or initial reagents in organic synthesis. Due to their high volatility, solubility in water (equal to about 1 g/l) and mobility, chloroalkanes and chloroalkenes are able to penetrate through the concrete walls of sewer systems and enter groundwater. The lipophilic nature of these contaminants results in their accumulation in adipose tissues of animal organisms and incorporation in food chains [162].

Tetrachloromethane is generally used for the synthesis of fluorine chlorohydrocarbons and fat solvents. It is estimated that 5–10% of all tetrachloromethane produced is spread in the environment. Under aerobic conditions (in the atmosphere and the superficial layers of reservoirs, rich in oxygen), the half-life of CCl_4 is 60–100 years but in the bottom layers of reservoirs, anaerobic microorganisms assimilate tetrachloromethane within 14–16 days.

The hazardous action of tetrachloromethane on human health arises from its possible transformations in the liver. Under the action of cytochrome P450-containing monooxygenase a chlorine atom is eliminated from tetrachloromethane followed by trichloromethyl radical formation (Fig. 1.2) [10]. The trichloromethyl radical is further transformed into chloroform by the splitting of a hydrogen atom from unsaturated fatty acids. By these reactions, lipoperoxidation is initiated, resulting in formation of radicals of fatty acids, which are then transformed into diene radicals. Further, diene radicals join oxygen and form hydroperoxides, which disintegrate and induce the splitting of cell membrane phospholipids. The damaged membrane negatively affects intracellular metabolic processes, such as the functioning of mitochondria, the Golgi apparatus and other intracellular organelles. The consequence is the destruction of the liver function, with the consequence that proteins (e.g. aminotransferases, α-globulins etc), biliary pigments (e.g. bilirubin), minerals and other substances penetrate from the liver into the blood, the electrolyte content of the blood is changed and toxic hepatitis develops [468].

As the number of substituted chlorines increases, the tendency of such compounds to form radicals (via monooxygenase action) increases and in turn, their hepatotoxic activity increases [162].

Trichloroethylene (TCE) belongs to those chlorinated hydrocarbons that have a poisonous effect on the liver. In industry TCE is mainly used to eliminate fat from metal surfaces, as a solvent for a number of compounds (including natural ones), and in the synthesis of organics. It has been estimated that about 90% of all produced TCE is dispersed in the environment, largely in the air, and the rest in solid wastes and sewage [162]. TCE is very stable under aerobic conditions. In seawater, its half-life is 90 weeks, and in fresh water it varies from 2.5 to 6 years. As a result of the action of anaerobic bacteria, the half-life is shortened to 40 days, but in the ground TCE may remain for several months. Poplar, cottonwood (*Populus* sp.), willow, clover (*Trifolium* sp.), alfalfa (*Medicago sativa*), rye (*Secale* sp.), sorghum (*Sorghum bicolor*) and other plants assimilate TCE and other chlorinated aliphatic compounds, degrading their carbon skeleton to CO_2 [151].

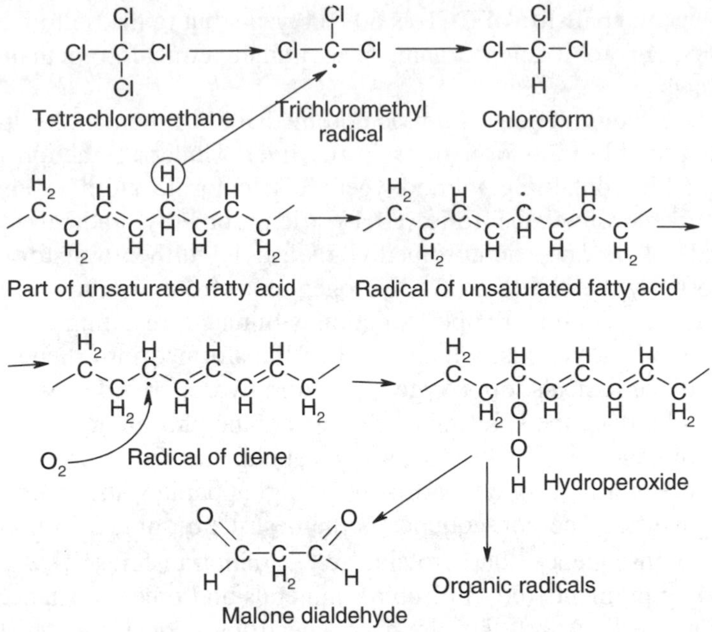

Fig. 1.2. Initiation of fatty acid peroxidation by trichloromethyl radical

The toxic effect of TCE on animal organisms is mediated by the action of a monooxygenase. Initially TCE is transformed into an epoxy compound, which in turn, is transformed into trichloroacetaldehyde [162]:

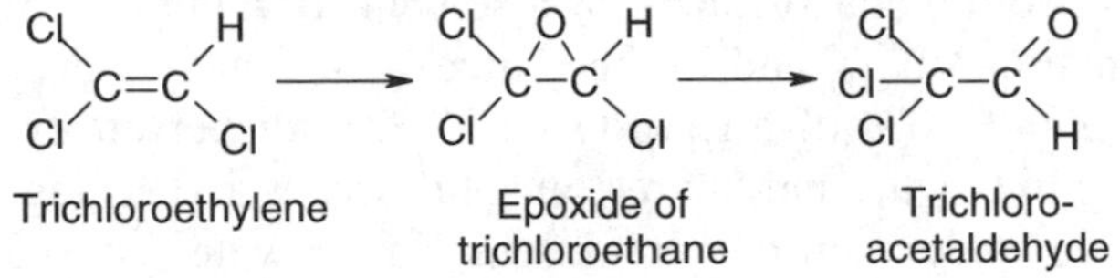

Trichloroethylene   Epoxide of trichloroethane   Trichloro-acetaldehyde

Fig. 1.3. Formation of trichloroacetaldehyde from trichloroethylene

Besides aldehyde, trichloroacetic acid, trichloroethanol and chloral hydrate can be formed in the organism. Trichloroaldehyde is a mutagenic compound, as it actively reacts with DNA.

Vinyl chloride, the monomer of polyvinyl chloride, PVC (the basis for the production of linoleum, washable wall-papers, leather substitue, plastic bottles and many other organic polymer products) has analogous carcinogenic characteristics [18].

1.1.9 Benzene and its homologues

Benzene and its homologues are extremely toxic. Nowadays over 90% of the benzene produced is connected with the petrochemical industry, the rest with the coal industry and natural gas. The United Kingdom (UK), the major exporter of benzene, annually produces one million tons of this substance. Since 1980 the production and use of benzene has been limited in the countries of the European Economic Community, the UK and the United States because of its high toxicity.

Benzene   Toluene   Ethylbenzene   Xylene

Most benzene and its homologues in different admixtures (the so-called BTEX – benzene-toluene-ethylbenzene-xylene) are used in many types of fuel to increase the octane rating, instead of the very toxic petrol additive tetraethyl lead. Besides, benzene is used as the raw material in the production of styrene, cyclohexane, ethylbenzene, cumene, nitrobenzene, aniline etc. and as a solvent or additive in the production of dyes, pesticides, inks, rubbers, glues, lubricants, spot removers, furniture waxes, detergents and drugs.

The main anthropogenic sources of benzene contamination are:
– Leakage of crude oil and petrochemicals during refining.
– Emission from coal tar distillation and coal processing plants.
– Industrial emissions where benzene is the final product or initial reagent in organic synthesis.
– Emission from burning oil and fossil fuels.
– Leakage from underground fuel tanks.

After production or use, benzene is initially distributed in the atmosphere, whence it penetrates into other ecosystems. Waters of oceans, seas, lakes, reservoirs, rivers, soil and sediments contain benzene in different amounts.

Benzene is detected even in space, although it is not there as a result of human activities. It is supposed that benzene evaporates from stars at a particular stage of their development. It is considered that benzene molecules are formed around carbon-rich old stars, e.g., red giants [71].

Benzene and its homologues are well-known carcinogenic substances that cause leukemia [540]. After entering into the liver or lungs, benzene, as a nonpolar and relatively stable compound, undergoes initial oxidation by a cytochrome P450-containing monooxygenase, forming benzene oxypin and benzene oxide [424]. These compounds are characterized by an increased solubility and reactivity compared to benzene.

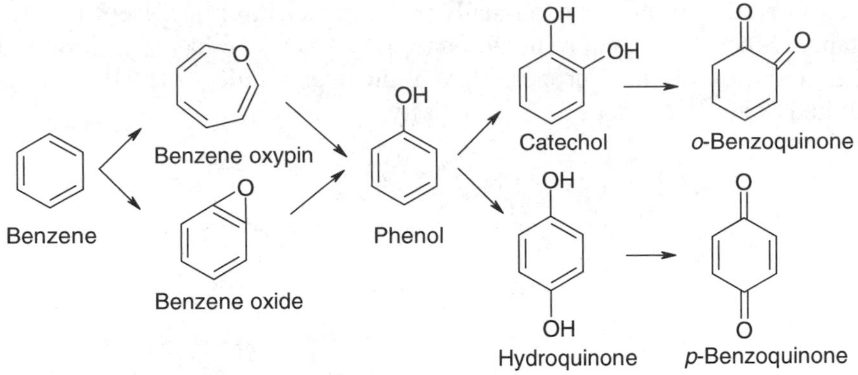

Fig. 1.4. Formation of benzoquinones from benzene

Furthermore, the products of the primary oxidation of benzene are moved from the liver to other tissues, including bone marrow, via the blood. Benzene oxypin and benzene oxide undergo further enzymatic transformations in these tissues: they are reduced to phenol that in turn is oxidized to catechol or hydroquinone. These diphenols are oxidized to benzoquinones. Transformation of diphenols is catalyzed preferentially by enzymes of marrow cells. The benzoquinones formed are characterized by an enhanced reactivity. Each of them can bind proteins or nucleic acids via oxo-groups, and this leads to the destruction of the normal biological functioning of the macromolecules [424].

1.1.10 Heavy metals

Heavy elements are defined as chemical elements whose density is at least five times heavier than that of water. Among 35 widely occurring metals, 23 are heavy elements or heavy metals: Ag, As, Au, Bi, Cd, Ce, Cr, Co, Cu, Fe, Ga, Hg, Mn, Ni, Pb, Pt, Te, Th, Sb, Sn, U, V and Zn [192]. In small amounts, most of these elements are indispensable for many organisms, but their enhanced doses induce acute or chronic poisoning. The toxicity of heavy metals is apparent in reducing growth and development in microorganisms and plants, and seriously harming the health of animals and humans. In particular, heavy metals may disrupt the normal function of the central nervous system and cause changes in the blood content, and adversely affect the function of lungs, kidneys, liver and other organs. The long-term action of heavy metals may cause the development of cancer, allergy, dystrophy, physical and neurological degenerative processes, Alzheimer's and Parkinson's diseases, etc.

All the above-mentioned determines the inclusion of some heavy metals in the list of the 20 most hazardous substances created by the ATSDR and EPA. The heavy elements arsenic, lead, mercury and cadmium, which pose serious ecological problems (occupying first, second, third and seventh positions respectively, in a global toxicity ranking according to the ATSDR and the EPA in 2003), will be described in the following section.

Arsenic

This element occurs in the ores of copper, lead, iron, nickel, cobalt and other metals. All compounds of arsenic are highly toxic. Upon heating they are decomposed, releasing poisonous arsenical vapors. The main sources of environmental contamination by arsenic are emissions from the mining industry, the production of arsenic and its compounds, the smelting of copper, lead and zinc, and the burning of coal.

Arsenic oxides, arsenites and arsenates are mainly used for preserving wood (88% of these compounds in the USA is used for this purpose [19]); arsenic compounds are also constituents of insecticides, herbicides and desiccants, and are used in the production of glasses, anticorrosive alloys, solders (a metal alloy used when melted to join or patch metal parts), ammunitions, and lead acid accumulators. High-purity arsenic is used in semiconductor devices, including solar batteries, light-emitting diodes, lasers and integrated circuits. Until the 1970s inorganic compounds of arsenic were used in medicine for the treatment of leukemia, psoriasis and asthma.

Volatile arsenic compounds distributed in the atmosphere accumulate on the surface of soils, in reservoirs and in plants and thereby enter food chains. A major ecological disaster occurred in Bangladesh through an ill-conceived project promoted by the WHO to "improve" the supply of drinking water for millions of peasants in the Ganges delta, who had previously relied on the river. Numerous boreholes were sunk, penetrating substrata rich in arsenic-bearing minerals. Several years elapsed, during which vast numbers of people daily imbibed considerable quantities of arsenic, before any harm was noticed. As a result, millions of people are now suffering from chronic arsenism.

Globally, however, arsenism, or arsenic poisoning, is a rare disease [487]. Chronic poisoning is also observed due to prolonged contact with arsenic vapor and dust, which is quite often lethal. Poisoning by nonlethal doses causes hemolysis of erythrocytes, damage to the liver and kidneys, skin irritation and deleterious effects on the central and peripheral nervous systems, and the digestive tract. Depending on dose, arsenic and its compounds are carcinogenic (International Agency for Research on Cancer) [246], causing cancer of the skin, liver, intestines, bladder and lungs. The long-term use of arsenic-containing insecticides against phylloxera induced the so-called "cancer of the viticulturist", a cancer disease typical for people working in vineyards [145].

Some tropical algae are resistant to arsenic. They can absorb arsenic as arsenate ions, reduce it to arsenite and bind it with phospholipids. The formed conjugates are stored in lipid droplets or cell membranes [145]. A few species of mycelial fungi and bacteria are capable of taking up and transforming arsenic compounds. For example, methanogenic bacteria can transform inorganic arsenic into methylated compounds under aerobic conditions, which are reduced by enzymes to volatile alkylarsines [162].

Lead

Lead is a widely used heavy metal. Metallic lead and its compounds (oxides, halogenides, carbonates, chromates, sulfates, etc.) are used in the production of accumulators, piezoelectric elements, gums, glasses, glaze, enamel, drying oil and putty; in polygraphy for the production of dyes and pigments; in lacquers and dyes to enhance opacity; as an antiknock additive in petrol; and for protection against γ-radiation; etc.

World lead production is several million tons annually. The most important anthropogenic sources of lead are waste extracted during high temperature processes in metallurgy, metal working, engineering, chemical, chemico-pharmaceutical, petrochemical and other industries; and the exhaust gases of internal combustion engines using lead-containing petrol. High levels of lead contamination are typically found in the soils of military proving grounds [1].

Lead compounds (oxides, chlorides, fluorides, nitrates, sulfates, etc.) are emitted as solid particles from internal combustion engines together with the exhaust gases. The cultivation of agricultural plants, especially fast-growing vegetables near roads is, therefore, not recommended. In recent years, legislation has greatly diminished the use of lead as an anti-knock agent in motor fuel in many countries, and lead pollution near roads has correspondingly diminished. Instead of lead compounds, organic compounds such as methyl *tert*-butyl ether (MTBE) are added, which may be carcinogenic and have already reached appreciable concentrations in many freshwater lakes and reservoirs.

Excess lead in the soil decreases soil microbiocoenoses. The degree of toxicity caused by lead to the microflora depends on the type of soil. In black earth the neutralization of toxicity is faster than in other soils [418]. Some species of eukaryotes (microscopic fungi) and prokaryotes (bacteria) are rather resistant to lead compounds. Actinomycetes and bacteria that assimilate molecular nitrogen are more sensitive to lead than the representatives of other taxonomic groups of microorganisms. Hence, they can be used as bioindicators for the level of lead contamination.

A soil concentration of lead that decreases the harvest or plant height by 5–10% is considered as toxic. When the lead content in soil is > 50 mg/kg, its concentration in vegetable crops exceeds the permissible level. In humans, about 90% of the lead enters through food, and 60–70% of the lead is of plant origin [418].

Lead is a moderate toxicant. In humans it causes chronic poisoning (saturnism) with varied clinical symptoms, including damage to the central and peripheral nervous systems, bone marrow, and blood vessels, suppression of protein synthesis, and perturbation of the cell genetic apparatus. It also causes gonado- and embryotoxicities and activation of oncological processes [85].

All lead compounds are characterized by a similar action, with differences in their toxicities being explained by their different solubilities in the gastric juice, intestinal fluid, blood and cytoplasm. Sparingly soluble lead compounds also undergo transformations in the intestines, their intermediates being characterized by increased solubility and absorptivity. White lead, and the sulfate and oxide of bivalent lead, are more toxic than other compounds. Lead compounds with a toxic anion, such as ortho-arsenate, chromate and azide, are especially toxic. Organo-lead compounds, especially tetraethyl lead, used for enhancing the octane rating of petrol, are biocides. Volatile tetraethyl lead rapidly evaporates in air, and is split into radicals under the action of UV light (250 nm). Radicals of triethyl lead react with acceptor substances (A, see Fig. 1.5) [162].

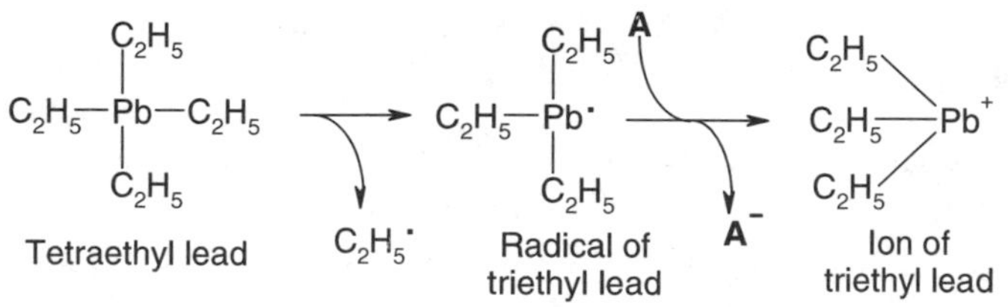

Fig. 1.5. Formation of triethyl lead cation from tetraethyl lead

The $Pb(C_2H_5)_3^+$ is hydrophilic due to its electrostatic charge, and the ethyl groups impart a lipophilic character onto the ion. Hence the triethyl lead ion can easily penetrate cell membranes, and thereafter bind with the sulfur atoms of proteins and peptides, causing structural changes.

Mercury

Mercury is present in the earth's crust in the form of cinnabar (HgS), a relatively harmless substance. However, natural processes and human activities have led to the accumulation in the oceans of more than 50 million tons of the toxic compounds of this heavy metal. Natural mechanisms of mercury dispersal are aerial degradation of rocks and volcanic emissions. The main anthropogenic sources are the fossil fuel-based power plants, waste incineration, electrochemical production of chlorine and mercury-containing devices, marine paints, pesticides, pharmacological preparations, and catalysts for the synthesis of organics [198].

Under natural conditions mercury compounds are mainly adsorbed on fluvial sediments, but the mercury is then slowly released and dissolves in water as the Hg^{2+} ion. Under the action of anaerobic microorganisms it rapidly interacts with organic substances and forms the extremely toxic compound dimethyl mercury ($CH_3-Hg-CH_3$) and the methyl mercury (CH_3-Hg)$^+$ ion [145]. With its relatively high solubility, methyl mercury rapidly penetrates aquatic organisms (algae, shellfish, fish, etc.) and thus enters into the food chain. Methyl mercury is especially hazardous for animals and humans, as it quickly penetrates from blood into cerebral tissue, damaging the cerebellum and cortex. Clinical symptoms of such destruction are torpor, loss of orientation and bad vision. Mercury poisoning may be ultimately lethal [407].

Mercury compounds specifically inactivate some enzymes, particularly the cytochrome oxidases participating in the respiratory process. Besides, mercury can bind SH-groups, inhibiting SH-group-containing enzymes, structurally altering proteins, and destroying cell membranes and DNA.

Cadmium

Cadmium, characterized by high mobility and permeability, is an extremely toxic heavy metal. Metallic cadmium and its compounds are mainly used in the production of pigments for stabilizing plastics (especially polyvinyl chloride), as well as in the production of accumulators, control rods for nuclear reactors, electric cables, motor radiators, solders, alloys, fertilizers etc.

Cadmium sulfide (CdS) and cadmium selenide (CdSe) are heat-stable yellow and red pigments used in polygraphy, for the production of lacquers, dyes and rubber goods, and the painting of leather. Cadmium oxide (CdO) and carbonate $(CdCO_3)$ are used in the coloring of glass, in the production of enamel, and the glazing of ceramics.

The most important anthropogenic sources of cadmium emission into the atmosphere are the production of steel and other metals and the burning of fossil fuels and garbage. Contamination of soil and water arises from fertilizers and sewage from industrial plants [277].

Cadmium is typically bound to dust particles, which can penetrate into organisms when breathing. Plants contact cadmium during its precipitation from the atmosphere when it penetrates into leaves via their cuticles. Cadmium accumulation in plants causes perturbation of normal growth. On the other hand, many species of fungi are able to accumulate high concentrations of cadmium while retaining full vitality [277].

The main source of cadmium in animals is food. Cadmium reduces the activity of digestive tract enzymes such as trypsin and pepsin. Besides, cadmium antagonizes calcium: calcium deficiency leads to an accumulation of cadmium in the bones. Young animals have a higher calcium requirement than adults, therefore they accumulate cadmium in higher amounts. Enhanced accumulation of cadmium induces the disease itai-itai, revealed by a reduction of the calcium content of the bones, leading to osteomalacia [526]. In kidneys, liver and the gall bladder, cadmium binds with proteins and peptides forming metallothioneins, which participate in the exchange of cadmium among different tissues and organs [358]. The most sensitive and easily damaged organ is the kidney. Excess cadmium inhibits the action of zinc-containing enzymes and damages the normal functioning of the kidneys, resulting in proteinuria. In the liver, cadmium blocks the activities of enzymes containing SH-groups [315].

Earthworms are capable to rapidly accumulate cadmium from soil; consequently they are often used as bioindicators of cadmium [277].

1.1.11 Gaseous contaminants

Many gases contained in air are hazardous ecocontaminants when present above their natural concentrations, and, as such, can cause serious pollution of the environment. Oxides of carbon, nitrogen, sulfur, hydrogen sulfide, methane, chlorofluorocarbons, volatile organic compounds (VOCs) etc. belong to the class of potentially hazardous compounds.

Carbon monoxide (CO)

Among the gases playing a special role in the contamination of air, carbon monoxide formed by the incomplete burning of carbon-containing substrates should be considered as one of the most poisonous. The annual global emissions of CO into the atmosphere have been estimated to be as high as 2600 million tons, of which about 60% are from human activities and 40% from natural processes [149]. The latter are mainly volcanic activities and the photochemical oxidation of methane in the atmosphere. Other significant sources of CO are exhaust gas emissions. The optimum condition for fuel oxidation in internal combustion engines is reached only within a definite, and fairly narrow, working regime corresponding to about 75% of the engine capacity, but under other conditions such as idling and at the initiation of combustion, the CO content in the exhaust gases is significantly increased. Special catalysts providing full oxidation of the fuel to CO_2 have been made to purge the exhaust gases from CO. Gases emitted by automobile engines do not contain CO in a great amount, but in cities and areas of increased atmospheric pressure, or temperature inversion, CO may attain dangerous concentrations. The ambient concentrations measured in urban areas depend greatly on the density of the combustion-engine-powered vehicles, and are influenced by the topography and weather conditions. The carbon monoxide concentration in streets varies greatly depending on the distance from the traffic [417].

CO has a deleterious action on humans for many reasons. The most important reason is that CO competes with oxygen in binding to hemoglobin in the blood, which causes a sharp reduction in the oxygen-carrying capacity of the blood. The affinity of hemoglobin towards CO is 200–300 times higher than towards O_2. A CO concentration in the atmosphere equal to 0.006% is high enough to bind half of the blood hemoglobin [162]. Besides, CO can form highly toxic compounds, the carbonyls.

The increase in CO concentration in the air due to the continuous anthropogenic and natural emissions of CO and the relative stability of CO in the atmosphere is counteracted by higher plants, algae and soil microorganisms that fix CO. Higher plants and microorganisms bind CO via sulfur-containing amino acids, or oxidize CO to CO_2.

Carbon dioxide (CO₂)

Carbon dioxide is the end product of the complete oxidation of carbon-containing compounds. Atmospheric CO_2 is in permanent exchange with soil, water and most organisms. The natural sources of CO_2 formation are: volcanic eruptions, exposure of carbon-containing rocks, rotting of organic compounds (microbiological decomposition), respiratory processes and forest fires. Undoubtedly, the amount of CO_2 released from all these sources would be sufficient to exterminate all organisms if CO_2 fixation in the natural world ceased. Higher plants permanently fixing CO_2 via photosynthesis are distinguished by their high CO_2 exchange potential. Photosynthesis and dissolution of CO_2 in seawater decreases the amount of CO_2 in the air to a level that does not inhibit vital processes.

Dry land and the oceans keep the processes of CO_2 emission and fixation in equilibrium. However, only a small part of the total carbon mass participates in this natural metabolic equilibrium. The enormous increase in the amount of burnt fuel has led to the increase in the CO_2 content in the atmosphere. Other reasons for the increase of CO_2 content may be a decrease of the ability of soils to fix CO_2 and deforestation, especially of tropical rain forests. The balance between fixing and emitting of carbon has thereby been significantly changed.

CO_2 circulates in the atmosphere for typically 2–4 years [162]. The deleterious effect of increased CO_2 levels is expressed not only in its immediate toxic action on many organisms, but also in its high absorption of infrared (IR) rays. As the sun heats the earth's crust part of the heat returns to space as IR radiation. This reflected heat is partially captured by IR radiation-absorbing gases, including CO_2, that are heated as a result. This is the so-called "greenhouse effect", leading to global warming [38]. In an effort to avert these climatic changes, the leading countries of the world, signed the Kyoto protocol in 1992, designed to reduce the emission of CO_2 into the atmosphere. Various other attempts to diminish atmospheric CO_2 are being considered. For example, huge amounts of CO_2 have been pumped into rock formations from which natural gas had been extracted, earning the involved businesses "carbon credits" [360].

Sulfur dioxide (SO₂)

The natural sources of SO_2 are volcanoes, forest fires, sea foam and the microbiological transformations of sulfur-containing compounds. Once in the atmosphere, SO_2 can bind with water, thereby maintaining its stable concentration in the air.

The burning of coal and oil, metallurgical processes and the processing of sulfur-containing ore form anthropogenic SO_2. Most of the anthropogenic sources of SO_2 (about 87%) are connected with power engineering and industry. The total anthropogenic SO_2 comprises more than 90% of the SO_2 existing in nature [150].

SO$_2$ resides in the atmosphere for approximately two weeks. This interval is too short for its global dispersion. Therefore, in neighboring geographic regions, differences in the SO$_2$ content of the atmosphere are observed. The problem of SO$_2$ pollution is more or less typical for highly developed industrial countries and their neighbors.

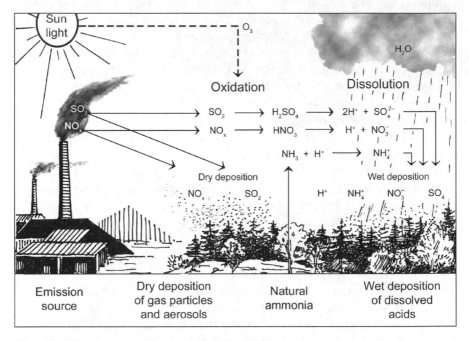

Fig. 1.6. Formation of acid rain [381]

In the atmosphere SO$_2$ (together with nitrogen oxides (NO$_x$), see below) undergo chemical transformations, the most important among which are oxidation and acid formation, which leads to formation of so-called "acid rain" (Fig. 1.6). These reactions proceed in the presence of UV radiation, oxygen or ozone.

It is estimated that 60–70% of acid rain is caused by sulfur dioxide. SO$_2$ and acidic precipitation induce corrosion of metalware and organic materials such as leather, paper, clothes, rubber and dyes. SO$_2$ has a toxic action on organisms. It is harmful for photosynthesizing plants. Hydrosulfite ions (HSO$_3^-$) are especially toxic for plants. They react with peroxides of unsaturated fatty acid phospholipids by forming radicals and thereby destroying biomembranes [162] according to the following reaction:

$$HSO_3^- + RCOOH \rightarrow HSO_3^{\cdot} + RCO^{\cdot} + OH^{-}$$

The $HSO_3^{\bullet}$ and $RCO^{\bullet}$ radicals disrupt the chloroplast membranes and oxidize and decolorize chlorophyll. Besides, the products of SO_2 transformation promote acidification of the cytoplasm, which results in the removal of the magnesium ion from the porphyrin ring of chlorophyll. HSO_3^{-} ions inhibit enzymes participating in the Calvin cycle during the photochemical fixation of CO_2. Therefore, leaves turn yellow and lose their photosynthetic potential as a result of exposure to SO_2. Sulfur dioxide also diminishes the transport of substances across membranes, leading to necrosis of leaves.

Nitrogen oxides (NO_x)

Nitrogen oxides are active contaminants of the atmosphere. The occurence of NO_x in nature is connected with electric discharges forming firstly nitrogen monoxide (NO) followed by nitrogen dioxide (NO_2). NO_2 can be released in small amounts in the process of silage fermentation. Oxygen-deficient soils host the microbiological denitrification of nitrates:

$$NO_3^{-} \rightarrow NO_2^{-} \rightarrow NO \rightarrow N_2O \rightarrow N_2$$

A high concentration of nitrate ion (NO_3^{-}) prevents the transformation of nitrous oxide (N_2O) into N_2, promoting the release of N_2O. Therefore, more than half the NO_x emitted from soil consists of N_2O.

Anthropogenic NO_x mainly consists of NO and NO_2 formed by the burning of fuels, especially at temperatures above 1000 °C. NO_x are also formed in the processes of nitriding, during the production of superphosphate, in the purification of metals by nitric acid, and in the production of explosives. The main source of NO_x release is automobile transport. The tendency towards a more rational use of fuel via its complete combustion also, unfortunately, leads to an increase of the amount of released NO_x (the efficiency of an engine increases with increasing working temperature). Anthropogenic contamination by NO_x typically exceeds the critical level in areas with a dense population [248].

Nitrogen monoxide and dioxide participate in some photochemical reactions by forming ozone and peroxyacetylnitrate ($CH_3COO_2NO_2$) contained in smog. Smog formation will be discussed in the next chapter.

NO does not irritate the respiratory tract, and, therefore, humans can not sense it. If inhaled, NO forms an unstable nitroso-containing compound with hemoglobin, that rapidly transforms into met-hemoglobin. The Fe^{3+} ion of met-hemoglobin cannot reversibly bind O_2 and is, therefore, not available for oxygen transport. A met-hemoglobin concentration in blood equal to 60–70% is considered to be lethal.

Further away from the source of emission, NO is transformed into NO_2. This yellow-brown gas strongly irritates mucous membranes. Upon contact with moisture in the animal nitrous and nitric acids are formed, corroding the alveolar walls of the lungs. The walls become permeable and allow blood serum to pass into the lung cavity. Inspired air dissolves in this serum that blocks the gas exchange.

In the atmosphere NO_x are frequently encountered together with ozone. The action of ozone on the animal is similar to that of NO_2. Ozone also induces edema of the lungs, and blocks the normal movement of ciliary hairs in the bronchi that remove the xenobiotic. This damage creates favorable conditions for the onset of cancer. NO_x and ozone are hazardous for human health, even below the critical dose [378].

NO_x can contact plants directly via the atmosphere or acid rain, and indirectly affect them via photochemical transformations of oxidizers, like ozone and peroxyacetylnitrate. Acid rain containing NO_x seriously harms plants in ways similar to SO_2. Direct plant contact with NO_x rapidly results in the yellowing or browning of leaves, and the needles of conifers. The reason for these color changes is the transformation of chlorophyll a and b in pheophytins and the destruction of carotenoids. This process is induced by the hydroperoxide derivatives or radicals of fatty acids. Oxidation of fatty acids and the concomitant oxidation of chlorophyll lead to the disruption of membranes and necrosis. Fatty acids can be oxidized by the immediate action of NO_2, splitting off hydrogen by radical formation (Fig. 1.7) [162].

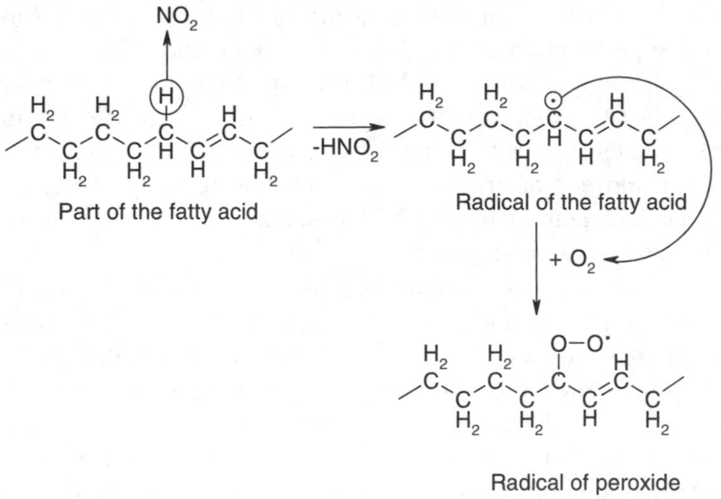

Fig. 1.7. Formation of peroxide radical from a fatty acid by NO_2 action

Besides, NO_2 can directly bind to double bonds, forming highly active radicals.

Nitrous acid formed in cells possibly enacts its mutagenic action via an oxidative deamination of DNA. The transformation of cytosine to uracil serves as an example (Fig. 1.8) [162].

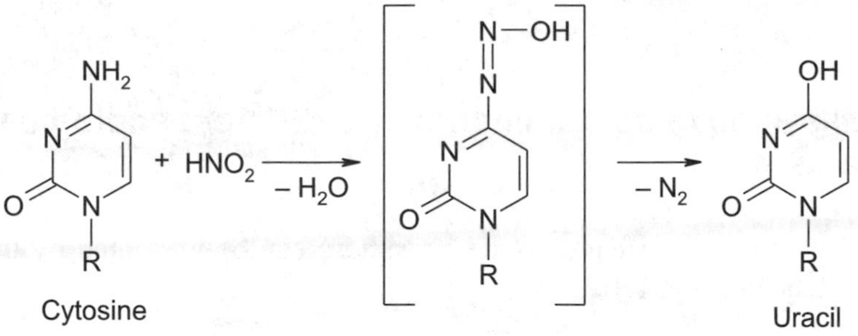

Fig. 1.8. Transformation of cytosine into uracil by the action of nitrous acid

Ozone

Ozone has a more toxic effect on higher plants than NO_x. It changes the structure of cell membranes, enhancing their permeability to water and glucose. As a result of these processes, leaf cell necrosis occurs, leading to a plant disease called silver leaf stain. This pathological condition damages the processes of transformation of assimilated substances, which then accumulate in cells, interrupting photosynthesis. Particularly, electrons excited by light form superoxide radicals, oxidizing ascorbic acid or forming hydrogen peroxide instead of reducing the nicotinamide coenzyme NADP. On one of the intermediate transporters of electrons (ferredoxin), hydroxyl radicals are formed initiating lipid peroxidation. Finally, the radicals oxidize chlorophyll and the leaves are bleached [360].

The hydroxyl radicals formed via the action of ozone react with the wax layer of the leaf surface or the needles of trees and cause their cracking. Microorganisms can easily penetrate into the cracks. The infection may result in deterioration of the entire tree. Peroxyacetyl nitrate also has a harmful effect on plants. This toxicant is photolytically dissociated into nitrogen dioxide and the peroxyacetyl radical. The latter inhibits chlorophyll and, hence, arrests the functioning of the photosynthetic apparatus.

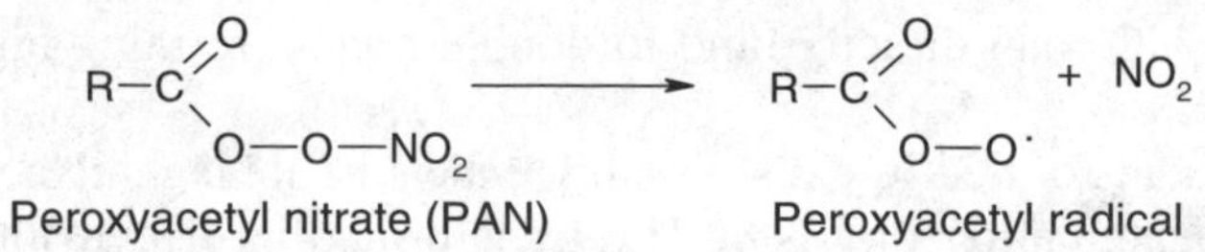

R–C$\overset{\displaystyle O}{\underset{\displaystyle O-O-NO_2}{\big\langle}}$ $\longrightarrow$ R–C$\overset{\displaystyle O}{\underset{\displaystyle O-O\cdot}{\big\langle}}$ + NO$_2$

Peroxyacetyl nitrate (PAN)   Peroxyacetyl radical

Fig. 1.9. Decomposition of peroxyacetyl nitrate with formation of peroxy-acetyl radical

1.2 Migration of contaminants into different ecological systems

The circulation of toxicants in the ecosphere can be presented by the following scheme (Fig. 1.10).

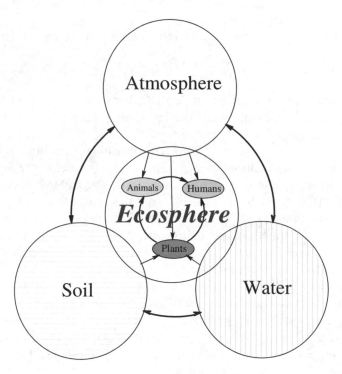

Fig. 1.10. The circulation of contaminants in the ecosphere [277]

Migration of contaminants is first of all the tendency of compounds of different structures and molecular masses to be dispersed in the environment. It is a very complicated multistage process controlled by many physical, chemical and biological mechanisms and factors, in particular:
- Basic physico-chemical characteristics of substances such as molecular mass, water solubility, "hydrophobicity" (calculated as the coefficient of substrate partitioning between nonpolar and polar solvents – typically n-octanol and water – designated as K_{ow}); vapor pressure (the determinant of substrate volatility); the presence of reactive functional groups, etc.
- Physical processes of mass transfer of substances, such as adsorption, desorption, diffusion, impedance, convection, dispersion, dry and wet precipitation, etc.
- Chemical processes such as oxidation, hydrolysis, photolysis, conjugation of toxic compounds or their derivatives with natural raw materials, etc.
- Geographical processes of substance circulation such as wind, precipitation, and ocean flow, river transportation, etc.
- Biological activities of organisms participating in the global processes of substance circulation in nature – bioconcentration, biomultiplication, bioaccumulation, biotransformation, biodegradation, biotic transportation of substances, etc.

The initial stage of toxic compound dispersion is escape from the zone of initial occurrence. The rate of this process depends on the technology associated with the handling of the chemicals (for instance, in the case of pesticide application, it is very important to know how it was dispersed – from the soil or from an airplane). Much depends on the closeness of the application system to the desired recipient, the topography, etc. One of the key features of toxic compounds determining their distribution is fugacity [323], i.e. the tendency of substances to come out of the phase in which they exist. The initial stage of substance dissemination in the zone of their initial application is followed by their nonhomogeneous distribution in the nearest ecosystem. Biotic and abiotic transfer of substances in the natural environment – soil, water and air – is an important part of this stage of movement and distribution of environmental contaminants [433].

1.2.1 Migration of contaminants between soil and water

Contaminants are often dispersed in the soil via rainwater or artificial irrigation. On the other hand, water flowing on the surface of soil (so-called runoff water), or penetrating into the depths of the soil, could lead to significant contamination of groundwater by soil pollutants. Therefore, it is obvious that the basic processes leading to chemical contamination of soil and groundwater by toxic compounds takes place at the soil-water phase boundaries [277].

The processes of adsorption are of utmost importance in the chemical contamination of soil and the dissemination of contaminants. Due to the different adsorption abilities of soil components, an uneven distribution of contaminants takes place in the soil. Contaminants are adsorbed on lipophilic soil organic matter, mineral (clay) particles, and covalently bound with humus components. The adsorption process is typically not completely reversible, since an opposing process – desorption – proceeds in salty aqueous solutions, which is however incapable of releasing the contaminant molecules bound chemically or electrochemically to humus macromolecules and clay particles from the soil [240].

Adsorption significantly limits mass transfer and hence distribution in the soil. The process of mass transfer is the basic power of contaminant migration in soil and is accomplished via diffusion, convection and dispersion of substances [26].

Diffusive mass transfer, i.e., the spontaneous movement of structural particles of a substance (molecules, atoms or ions) along a concentration gradient is accomplished as the result of Brownian (thermal) movement. The intensity of this process does not depend on the water flow velocity and is determined by the following parameters [277]:
– Soil porosity (soil pore sizes and quantities).
– Molecular characteristics of diffusing substances (molecular mass and volume).
– The difference in the concentrations of the particular compound in soil and water (magnitude of the concentration gradient vector).

Hence, high soil porosity, large molecules (contaminants), and a small concentration gradient, etc. are factors reducing diffusion. The sum of these factors is called the impedance; its value indicates how slowly diffusion proceeds in soil in comparison with diffusion in free liquids.

Besides diffusion, contaminant migration is forced by the flow of dissolved substances inherent to the flow of their solvent (water), i.e., convective mass transfer. Its velocity depends on the volumetric water flow and the concentrations of the dissolved substances.

Heterogeneity of soil pores promotes dissimilar movement of flowing compounds in soil pores, causing dispersive mass transfer. Dispersion plays an important role in the total mass transfer of chemical substances in groundwater.

More complicated processes of mass transfer due to many different soil parameters include the following: (1) the capillary water rise in soil caused by water evaporation from the soil surface; (2) the existence of a triple phase balance between the contaminant, its dissolved phase and its adsorbed phase; (3) the water flow from the soil macro pores (fissures, paths of earthworms, etc.), etc.

When discussing the process of toxic compound migration from soil to water, it is very important to keep in mind that contaminants in most cases are accompanied by the products of their partial transformation, mainly caused by the action of the enzymes of soil microorganisms and of plant exudates [277]. Intermediates could also form due to abiotic reactions proceeding under the action of solar irradiance, atmospheric oxygen and water. Soil mineral substances (iron, aluminum, etc. oxides) often serve as catalysts for such transformations.

1.2.2 Migration of contaminants between water and air

The partitioning of environmental contaminants between water and air phases is a dynamic process that occurs in both directions. Both processes of chemical compound transfer from water solution into the atmosphere (volatilization) and in the opposite direction (precipitation from air into water) are diffusion-controlled and subject to the same rules, but in the opposite sense [313].

The velocity of transfer of a chemical compound through the water–air boundary is directly proportional to the difference in their concentrations in the two phases. The flow of each substance is directed towards the decrease of its concentration. In chemically polluted water, the contaminant concentration decreases exponentially with time. This is due to the fact that air is a more open system than water, so under natural conditions the concentration of substances in the gas phase is considerably lower than in the water phase.

The fugacity of contaminants in the water–air system completely depends on their volatility, which is determined by many parameters, viz. substance transport velocity in the liquid and gas phases, temperature and Henry's constant [114]. This last parameter quantifies the correlation of the mean substance concentration in the gas and water phases. The total substance transport velocity characterizing substance volatility is equal to [277]:

$$K = \left[\frac{1}{k_1} + \frac{RT}{k_g H} \right]^{-1} \tag{1.1}$$

K is the total velocity of substance flow, R is the universal gas constant, T the absolute temperature, k_1 the substance transport velocity in the liquid phase, k_g the substance transport velocity in the gas phase, and H is Henry's constant.

From equation 1.1 it follows that the value of Henry's constant is an important factor determining the process of volatilization of a substance in the liquid phase. For instance, if the value of Henry's constant is high, which is characteristic of substances with a high vapor pressure or low solubility or both (benzene, toluene, organochlorine compounds and many other contaminants) the right hand term (RT/k_gH) in the velocity equation is far less than the left one $(1/k_1)$. In such cases the substance transport velocity generally depends on the value of k_1. This means that the volatility of a substance is controlled by the liquid phase and depends on k_1. In turn k_1 is conditioned by the molecular mass of the substance, its molar volume and exogenous conditions such as temperature, liquid turbulence, viscosity, and surface tension, and wind velocity, etc. If the value of Henry's constant is high (> 500 Pam^3/mol) they will move from water to air at almost the same velocity, despite differences in their physico-chemical characteristics, and even if their vapor pressures differ by an order of magnitude. For compounds with low values of Henry's constant (< 0.5 Pam^3/mol), characteristic of substances with a low vapor pressure, or high solubility in water, or both, it is the right hand term of the velocity equation that becomes more important, i.e. volatility is controlled by the gas phase [277, 460]. But, for relatively nonvolatile contaminants (or-ganophosphorus pesticides, surfactants) the transfer velocity from water to air is determined by the molecular mass of a substance, its molar volume, solubility and vapor pressure, as well as by exogenous factors (wind velocity, water turbulence etc.).

Besides the processes of substance evaporation from water, and dry pre-cipitation from the atmosphere into water, there are other pathways for the interchange of materials between these systems. Such pathways are, for in-stance, wind-driven seawater dispersal, and the release of contaminants from the atmosphere by precipitation (wet precipitation). The actual im-pact of the diverse transport pathways on the general interchange of chemical substances between water reservoirs and the atmosphere in many respects primarily depends on geographic location and local climatic con-ditions.

1.2.3 Migration of contaminants between soil and air

The processes of substance transfer between soil and air are most complicated, as factors controlling interchange between the three phases liquid–solid, liquid–gas and solid–gas [221] play an important role in the distribution of contaminants.

As in the case of mass transfer in the water–air system, the process of substance transport from soil to atmosphere (the so-called volatilization in the soil) is accomplished via diffusion. The velocity of volatilization depends on the molecular mass, temperature, pressure of the saturated vapor of the adsorbed substance and the velocity of its transport in the gas phase. Dry precipitation of chemical substances from the atmosphere onto soil follows the same rules as their volatilization from soil. Similar to the case of substance transport between liquid–air phases, the contribution of each direction of transport depends on the physico-chemical characteristics of the substance as well as the type of soil and climatic conditions [277].

Migration of chemical compounds adsorbed onto the dry soil surface from soil to air proceeds in a matter similar to migration in solution, i.e. following first-order reaction kinetics. Volatility from a wet soil surface is much higher than from a dry one. This phenomenon cannot be linked to co-evaporation of substances from water, because of the following reasons. Firstly, co-evaporation proceeds at higher temperatures and concentrations of contaminants than under natural conditions. Secondly, the reason for co-evaporation is interaction between the water and the evaporating substance (formation of hydrogen bonds, hydration, etc.), which is not characteristic for most contaminants. Thirdly, under conditions when the soil surface remains wet, the velocity of volatilization of many contaminants does not change when the water vapor rapidly saturates the surrounding atmosphere and hence its evaporation is rapidly suppressed. Therefore, evaporation of water and volatilization of chemical compounds from soil occur independently. The increase of volatility from wet soil in comparison with dry soil is mainly explained by partial desorption of chemical compounds during their displacement from water [476]. This points also to the fact that volatilization of chemical compounds from the wet soil surface takes place mainly in the liquid phase.

Environmental contaminants residing deep in the soil are transported towards the soil surface by different forces. Contaminants having a high Henry's constant (e.g., organochlorine solvents, the insecticides lindane and DDT) are transported from lower layers to the surface and their volatilization proceeds like that from water. For substances with a low Henry's constant (the triazine group herbicides, e.g. prometone), the upward transport occurs due to convection and capillary forces. This effect is called wicking [477]. The volatilization of chemical substances from the soil into the atmosphere depends additionally on other environmental conditions, such as soil type, temperature and wind speed. Another route of substance release from soil into the atmosphere is dust (wind erosion), and an alternative pathway of dry precipitation can be wet precipitation, i.e. transport of chemical contaminants from the air into the soil during atmospheric precipitation.

1.2.4 Geographical migration of contaminants

After distribution of toxic compounds between the air, water and soil as a result of the action of the physical factors discussed above, the so-called geographical transport of substances over various distances takes place. Geographical transport of substances is generally accomplished by currents of air and water [277].

Atmospheric transport is the most important pathway for the transportation and distribution of chemical contaminants into the environment. Substance transport in the air depends on the vertical structure of the atmosphere, as well as on the direction and velocity of prevailing winds. The mixing of substances with the air and their transport to the troposphere occurs rapidly, but exchange of the same substances between the troposphere and the stratosphere may last for several years. Tropospheric transfer of substances within a local area lasts several minutes; on a regional scale it takes from hours to several days; and for global migration from one day to several weeks is required [339].

The period of full mixing of toxic compounds with air in a terrestrial hemisphere varies from two weeks to three months, but transfer between hemispheres takes about a year as the equatorial zone of lower pressure around the globe considerably complicates air mass exchange between the north and south hemispheres.

Besides, atmospheric migration depends on meteorological conditions and the characteristics of the earth's surface. Predominant wind direction determine the direction of gaseous contaminant transfer while the height of the toxic gas distribution depends on wind velocity. With the increase of wind velocity, intermixing of gases with ambient air becomes more and more intensive, which leads to the dilution of contaminants. At the same time a high wind velocity complicates the ascent of toxic gases limiting their distribution in the upper layers of the atmosphere. Mountains, forests, interchange of low and high pressure regions, changes in the initial directions of wind as well as atmospheric precipitation mainly determine the horizontal distribution of water-soluble toxic gases. The temperature of the different (distinguishable) layers of the atmosphere also affects the direction of movement of harmful gases. In the troposphere the temperature of the air falls with an increase in height. Strong heating of the earth's surface promotes the formation of vertical air currents, carrying toxic industrial and transport emissions upwards. The temperature of these emissions often exceeds that of the ambient air, which also induces the ascent of toxic gases. Under such conditions the elevation of waste distribution reaches 500–700 m and more [162]. The velocity of vertical gas exchange is significantly reduced during inversion of the temperature, when for example the temperature does not decrease with increase in height but increases instead. Such a phenomenon could be the sudden nocturnal cooling of air layers closer to the earth's surface, or a superimposition of a current of warm air on cold lower layers. In the first case the thickness of the inversion layer reaches from ten to hundreds of meters, and the temperature increase varies from 0.1 to 15–20 °C. During the daytime and under strong sunshine conditions such a temperature difference rapidly disappears. Only in autumn and winter, when the earth is barely heated, can the temperature conditions described above be maintained during the day. In the second case the reason of the inversion is the heating of air compressed by an anticyclone in the upper layer of the troposphere. Inversion layers of warm air settle on cold elevated layers. As a result of the inversion, the gaseous contaminants accumulate under the lower inversion layer and significantly contaminate the airspace adjoining the earth. This is one of the reasons of smog formation [162].

The atmospheric circulation on the surface of the oceans induces sea currents, with an anticyclonic character at low latitudes and a cyclonic character at the middle and polar latitudes [277]. Under the influence of the trade winds strong trade currents directed to the west are formed, which along the eastern coasts of the continents change their direction and move alongside the coast up to a latitude of 40–50°. Winds dominating in the higher latitudes of the northern hemisphere change the direction of these streams in the opposite direction and form warm currents in the east (e.g. the Gulf Stream in the Atlantic Ocean and Kuroshio in the Pacific Ocean) [339]. The southern forkings (branching) of these currents join with the northern forkings (branches) of the trade currents, as weaker and colder currents move along the coasts again and, thus, the anticyclone circle becomes closed. In the Atlantic and Pacific Oceans such anticyclones are formed in the southern hemisphere. In the Indian Ocean the sea currents have a seasonal character and change the direction of the predominant winds (monsoons). Cold waters of the extreme north and south form several cyclone circles. The circulation of oceanic waters occurs in the vertical direction as a result of the convergence and divergence of the current, as well as by the convectional mixing of waters that occurs during the downward movement of waters cooled in winter. Besides, local winds, rising and falling tides, underground earthquakes, etc. may cause movement of water in the world's oceans. Oceanic circulation engenders the global transmission of chemical pollutants, which may last for several years. Water transport by the rivers is of essential value for regional distribution while superficial waters and groundwaters are the major factors involved in the local distribution of pollution.

1.2.5 Biotic migration of contaminants

Next to geographical transport, the biotic transfer of environmental contaminants by organisms also takes place in the biosphere. Generally, the scale of biotic transfer is smaller than that of the geographical. Biotic transfer occurs as a result of the active movement of insects, birds, fishes and animals accumulating ecopollutants (seasonal migrations of birds, the driving of cattle, etc.) and by the transfer of toxic compounds via the food chain. From the point of view of contaminant transfer, human activity may be regarded as biotic. The secondary transport of chemical compounds (pesticides, dioxins, PCBs, mercury, etc.), into soil and sediments by bioactivity is also of definite significance. It promotes the gradual transfer of contaminants into the liquid phase [277].

In spite of its relatively small scale, transfer via the food chain is prominent among the biotic processes of contaminant transfer. This process is one of the main routes of environmental contaminant into humans. In the XIX century, when mankind was less concerned about ecological problems, the possibility of toxic compound penetration through the food chain was anecdotally described by Alexandre Dumas. The hero of the novel, the Count of Monte Cristo, tells Mrs Vilford about a certain abbot Adelmonte, who feeds his rabbits with cabbage poisoned by arsenic and gives parts of the poisoned animals to his hens. One of the poisoned hens then falls prey to a hawk, which is also poisoned and falls into a fishpond, becoming food for pikes. The final victim of the villainous abbot is a man who eats fish for dinner. He is the last link in this fatal food chain.

Food chains represent relationships based on food sources between organisms, implying that each species serves as a food source for one or more other species. The classic example of contaminant penetration through a food chain is the transfer of DDT, starting with aquatic organisms [145], illustrated in Fig. 1.11.

For instance, if DDT is dispersed from an airplane over a stagnant reservoir infested with malaria mosquitoes, only an insignificant amount of insecticide is left in the water after several days. The DDT is completely transferred from the water into microorganisms, algae, or sediments, i.e., into the lower trophic levels of the food chain, with DDT accumulation proceeding up the food chain resulting in high toxic effects.

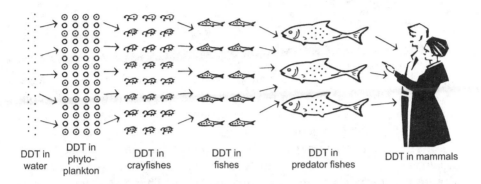

DDT in water — DDT in phytoplankton — DDT in crayfishes — DDT in fishes — DDT in predator fishes — DDT in mammals

Fig. 1.11. Insertion of DDT into the food chain [145]

Through the movement of contaminants to successive levels of a food chain, i.e. with the increase in trophic level, the concentration of a contaminant in the tissues of organisms may occur. Generally, the contaminant concentration per unit of biomass increases about tenfold at each successive stage of a food chain. This concentration takes place due to the following reasons [145]:

1. Once in the organism, toxic compounds that are persistent towards enzymatic degradation remain unchanged for a comperatively long time. Most toxic compounds tend to be barely metabolized, or do not participate in metabolism. As a rule, they are characterized by prolonged periods of biological degradation, and being lipophilic are accumulated in lipids and adipose tissues. If they are partially transformed, they bind with intracellular substances, especially with biopolymers. Therefore, their removal from the organism becomes practically impossible.

2. Toxic compounds concentrated in animals may effectively poison them. Hence, the animals become less mobile, lose reaction speed and are easy prey for other animals, at a higher trophic level. Contaminated animals are thus more likely to fall as prey than non-contaminated animals.

Accumulation of chemical toxicants in an organism, or bioaccumulation, occurs both as a result of feeding (biomultiplication) and due to penetration from the environment (bioconcentration) [277].

Bioconcentration is the main process of bioaccumulation by aquatic organisms. Toxic compound concentration in the organism is determined by the following equation:

$$C_A = \frac{k_1}{k_2} C_W = K_{BC} C_W \qquad (1.2)$$

C_A is the equilibrium concentration of the substance in the organism (µg/kg), C_W the concentration of the substance in water (µg/l), k_1 the consumption rate constant (day^{-1}), k_2 the isolation rate constant (day^{-1}), and K_{BC} the coefficient of bioconcentration [153].

As seen from equation 1.2, the content of toxic compounds in aquatic organisms depends not only on the concentration of those compounds, but also on the coefficient of bioconcentration (K_{BC}). This parameter indicates the relative rate of xenobiotic penetration into the organism in comparison with its removal. The K_{BC} values of the different chemical compounds in aquatic organisms may differ by up to 5 orders of magnitude and depend on the physico-chemical characteristics of the substance and lipids concentration in the organism. Therefore, the most important feature of a toxic compound is its lipophilicity, usually estimated by its n-octanol–water partition coefficient (K_{OW}). There is a linear relationship between the decadic logarithms of the coefficients K_{BC} and K_{OW}. Such a relationship indicates that the accumulation of lipophilic chemicals from water generally reflects their distribution between an aqueous phase and the lipid fractions of organisms, although bioconcentration of toxic compounds is also determined by other factors (e.g. the selective permeability of membranes, the intensity of specific biochemical reactions of binding and degradation of toxic compounds). Many experiments demonstrate that lipophilic toxic compounds, similarly to DDT, move very rapidly up the aquatic food chain [190, 346].

Unlike in aquatic organisms, the bioaccumulation of toxic compounds in terrestrial animals proceeds mainly by biomultiplication, i.e. orally via the digestive tract. The intensity of accumulation in these organisms also depends on the K_{OW} [277].

In higher plants the uptake of chemical contaminants takes place from the soil, water and air. The mechanisms involved in these processes differ from each other and will be discussed in a separate chapter.

1.2.6 Local contamination of ecosystems

Smog

The distribution of emitted chemical toxicants occurs much faster in the atmosphere than in the troposphere, therefore, accumulation of these harmful substances in air quickly follows their generation as a rule, but tends to be short-lived. However, some physico-chemical, geographical and meteorological factors can cause long-term local air contamination, e.g. the formation of smog.

Smog (the word is derived from the combination of **smo**ke + **fog**) is a mixture of gases forming a brownish-yellow or brown mist in large cities and industrial centers. There are two types of smog [162]:

1. Smog of the London type is a thick fog with an admixture of smoke and waste industrial gases. It is formed above the cities of northern latitudes in autumn and winter as a result of intense air contamination. The smog consists primarily of an aerosol, in which SO_2, H_2SO_4 and soot predominate.

2. Smog of the Los Angeles type is an aerosol with an elevated concentration of caustic gases, formed by the ultraviolet radiation of the sun engendering photochemical reactions in the gas exhausts from transport and industrial enterprises. The excessive use of air conditioners is also a significant contributor. This type of smog is also called "photochemical smog", and contains NO_x, ozone, peroxyacetylnitrate and several radicals. It is characteristic of southern cities in summer.

Smog formation occurs in regions where anthropogenic air contamination is strengthened by geographical features of the terrain, especially mountains that interfere with airflows, and meteorological conditions, such as temperature inversions in the troposphere, that obstruct vertical gas flow, thereby promoting the accumulation of air contaminants. Smog is usually observed under conditions of weak air turbulence (gently whirling airflows, or light winds). Smog reduces visibility, intensifies corrosion of metals and other materials, destroys vegetation, and heavily irritates the human respiratory tract. Intensive and prolonged smog is considered to be a reason for increased human morbidity and mortality.

Moist air and sulfur dioxide wastes promote the formation of smog of the London type in winter, when people heat their homes. Heating predominantly by burning of coal often leads to "pea soup fog". The visibility can be as little as half a meters. The process of this type of smog formation is as follows (Fig. 1.12):

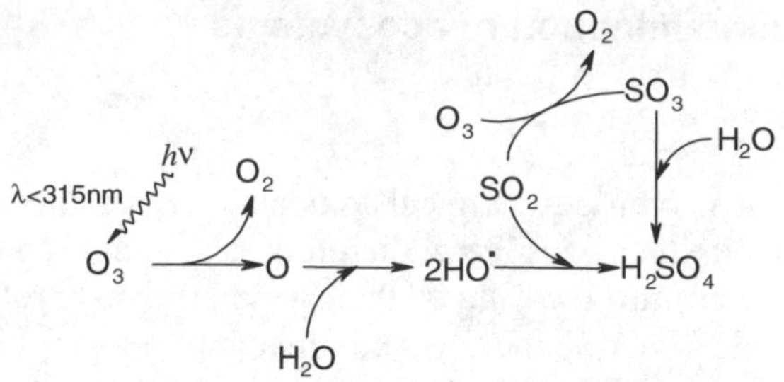

O₃ —hν λ<315nm→ O —O₂→ ... →2HO˙ →H₂SO₄

O₃ —→ SO₂ / SO₃, O₂, H₂O → H₂SO₄

H₂O

Fig. 1.12. Formation of sulfuric acid in London type smog

Under the action of UV light with a wavelength <315 nm, ozone de-composes and releases an excited oxygen atom in the singlet state, that together with atmospheric water vapor forms highly reactive hydroxyl radicals ($HO^{\bullet}$). These radicals easily oxidize SO_2 to sulfuric acid. The oxidation of SO_2 directly by ozone is also possible, leading to the formation of sulfur anhydride (SO_3) that actively attaches itself to water, also forming sulfuric acid. Finally, oxides and acids form a dense aerosol of sulfuric acid together with soot and water vapor.

Photochemical (Los Angeles) smog has a complex composition. It is a mixture of about a hundred toxic compounds including radicals with a very high oxidation potential [283]. The main sources of photochemical smog are NO_x and VOCs, such as ethane, propane, butane, ethylene, propylene, acetylene, methanol, formaldehyde, acetaldehyde, etc. [106].

Close to the ground, under the influence of sunlight with a wavelength <430 nm, NO_2 is split photolytically to NO and atomic oxygen. The latter then interacts with molecular oxygen on the surface of inert particles (Fig. 1.13; M: e.g. molecules of nitrogen) forming ozone. In air layers close to the ground ozone reacts rapidly again with NO, forming as initial products NO_2 and O_2 [467].

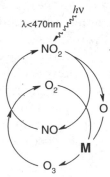

Fig. 1.13. Photolytic cycle of NO_2

Thus, a photolytic cycle is created and an equilibrium is established that prevents the accumulation of O_3.

Hydroxyl radicals, formed from ozone and water by rays of <315 nm forms CO_2 and hydrogen atom radicals (H˙) upon collision with CO. The hydrogen radicals can combine with molecular oxygen and the "inert" M particles, promoting the oxidation of NO to NO_2 via the intermediate of peroxide radicals (Fig. 1.14) [162, 277].

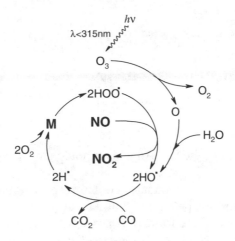

Fig. 1.14. Formation of NO_2 from NO and hydroxyl radicals

Thus, the photolytic cycle maintaining a stable concentration of ozone in the air is broken, and the accumulation of ozone begins. All reactions promoting the decrease in NO, or the increase of NO_2, or both, give a similar outcome.

In addition to ozone, photochemical smog also contains other secondary contaminants formed from primary contaminants – NO_x, CO, VOCs, etc. A simplified diagram of smog formation is presented in Fig. 1.15 [283]:

Hydroxyl radicals, formed from singlet oxygen and water, start to oxidize VOCs, getting into the air from different anthropogenic sources, including exhaust gases. At the first stage of oxidation, an alkyl radical is formed, that is converted into a peroxy radical by rapidly attaching itself to molecular oxygen. This leads to the oxidation of NO to NO_2, whereupon an alkyloxy radical is formed, that in turn is rapidly oxidized by oxygen forming an aldehyde. Hydroxyl radicals then again react by oxidizing the aldehyde to an acetyl radical. Then, a peroxyacetyl radical is formed as a result of the fast oxygenation of the acetyl radical [283]. The latter attaches itself to NO_2 and generates one of the main smog components, peroxyacetylnitrate, the toxicity of which has been discussed previously in section 1.1.11.

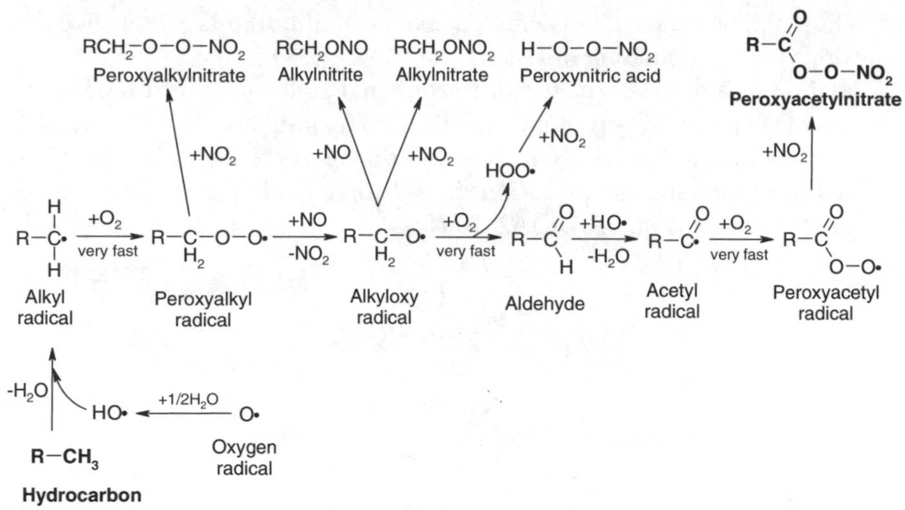

Fig. 1.15. Formation of the Los Angeles (photochemical) type smog

NO$_x$ also form other secondary contaminants in the photochemical smog. Among them are peroxyalkylnitrate, alkylnitrate, alkylnitrite, peroxynitric acid, etc., that are formed as a result of the reaction of NO$_x$ with intermediate radicals. Besides, all these radicals participate in the polymerization of olefins. As a result of hydrocarbon polymerization a smoky mass is formed, characteristic of this smog type, which generally significantly decreases visibility, and may be a nuisance to airplane pilots [283].

Oil contamination

In aquatic environments the dispersal of contaminants proceeds moderately rapidly. Not only reservoirs and rivers that collect wastewater are exposed to local contamination, also the seas and oceans. Contamination with oil hydrocarbons brings noticeable harm to marine ecosystems. In the middle of the last century the famous Norwegian scientist, traveller and writer, Thor Heyerdahl, while crossing the vast Pacific Ocean on his raft Kon-Tiki was surprised to observe many particles of coagulated mineral oil in the open ocean. He was the first to draw attention to this phenomenon.

According to data compiled by the Natural Resource Council Committee on Oil in the Sea [351] of the US National Academy of Sciences, about 1.3 million tons of oil and oil products enter world oceans annually. Oil penetrates into different ecological niches via the following pathways:

- Natural seeps from submarine deposits. This accounts for almost half the total oil contamination.
- Common tanker operations such as loading and unloading of oil. After unloading, the tanks are filled with seawater serving as ballast for stabilizing the tanker on its way back to the loading point. When the tanker is en route, salty seawater forms an emulsion with the oil product precipitates retained in the tanks after unloading. Upon arrival in the port of destination, the oil-containing ballast is dumped into the sea close to shore.
- Cleaning tankers, cisterns and reservoirs for oil and oil products.
- Tanker wrecks. For instance, following the shipwreck of the tanker Prestige in November 2002 in the Bay of Biscay, about 40 000 tons of crude oil poured into the ocean.
- Leakage from oil pipelines. Despite its high viscosity oil penetrates deep into the soil, reaching groundwater and spreading over long distances. For this reason oil is very often found in coastal swamps and seas.
- Loss of oil during drilling of oil wells located in the open sea.
- Rivers polluted with oil or oil products seeping into the riverbed.
- Wastes from the refining of crude oil.
- Jettisoning aviation fuel from aircraft.

Crude oil contains hundreds of different chemical components, the percentages of which vary over a wide range and depend on particularities of the oil deposits. Hydrocarbons constitute more than 75% of oil and the remainder is constituted by their derivatives, containing sulfur, nitrogen and oxygen. Oil hydrocarbons are divided into paraffins, cycloparaffins (naphthenes) and aromatic hydrocarbons.

Paraffins amount to 10–30% of total oil mass on average. The length of the paraffin chains can reach 43 carbon atoms (n-tritetracontane). Branched isomers of C_6–C_{25} alkanes, with short lateral chains, are more common.

Naphthenes comprise 30–60% of the oil mass. Most of them are monocyclic (cyclopentane and cyclohexane), but in some crude oils polycyclic naphthenes are also present.

The aromatic hydrocarbons benzene and naphthalene compose about 5% of the total oil mass, and the PAHs contained in highly boiled (340–400 °C) oil fractions make up a similar mass fraction. The fraction of naphthene-aromatic hydrocarbons containing both saturated and aromatic rings contain 9–25 carbon atoms and comprise 5–30% of the total oil mass. The residual oil fraction with a boiling temperature of >400 °C contains compounds with condensed heterocyclic rings connected to short paraffin chains.

After reaching the ocean, oil is distributed by wind action, ocean currents and tides. Due to its low density and hydrophobic nature spilled oil forms a thin slick on the water surface. All kinds of crude oil contain volatile components that rapidly evaporate. Approximately 25% of oil slicks evaporate within several days. Low molecular mass components are separated from the rest by dissolution. Aromatic hydrocarbons are characterized by higher solubility than paraffins. Components of low volatility are aggregated into oil clots that eventually precipitate onto the sea floor [485].

The oil slick content constantly changes due to the decomposition and transformation of its components. Oil hydrocarbons undergo degradation as a result of chemical oxidation and photooxidation, but in seawater they can be degraded by halophilic microorganisms. There is not one single microorganism capable of assimilating all crude oil components. Paraffin hydrocarbons are more easily decomposed microbiologically than cycloparaffins and aromatic hydrocarbons. The action of microorganisms is characterized by a high specificity, and hence the full decomposition of all oil components requires the participation of an entire consortium of microorganisms from different taxonomic groups, that is usually not present in any one oil-contaminated marine location. As a result of the partial microbial decomposition, oil hydrocarbons form a number of intermediate products that are more soluble in salt water than the initial hydrocarbons. These intermediates (alcohols, aldehydes, etc.) are often characterized by increased toxicity and are undoubtedly dangerous for sea organisms. The most significant factors controlling microbiological oil decomposition are temperature, nutrient availability and oxygen concentration in seawater. Full oxidation of 1 litre of crude oil requires the amount of oxygen contained in 375 tons of seawater [485].

Heavy undecomposable and non-precipitable oil residues appear on the water surface as floating resinous clumps, and end up in the coastal zones of the seas and oceans.

Oil components may affect ocean ecosystems for a long time. Even a small amount of oil may seriously damage the diverse biota of lakes and oceans (water containing even 1 ppm of oil becomes unfit for drinking). Oil patches on the surface of open water form an oil-water emulsion layer that partially blocks gas exchange between water and air. Under such anaerobic conditions, all organisms gradually suffocate by acidosis. Acidosis occurs when during breathing CO_2 accumulates in cells.

The primary harmful action of oil pollution at sea results from the direct covering of organisms with oil products. This causes the death of macroalgae, plankton and birds present in the tidal zone. Birds become especially vulnerable when their feathers are covered by oil. They lose not only the ability to fly, but also the thermal insulating properties of their feathers and the birds die as a result of overcooling. Moreover, they are poisoned by the oil swallowed during the preening of the feathers. Furthermore, the contamination of eggs can lead to damage of the embryo.

Water-soluble aromatic and polycyclic hydrocarbons, which easily penetrate into marine organisms, are especially harmful. The toxic effects of these hydrocarbons were discussed in the previous chapter. These toxicants can induce undesirable changes in sea organisms even at very low concentrations ($10^{-7}\%$). At concentrations of 10^{-6}–$10^{-5}\%$ serious destruction of physiological activity is observed, and 10^{-4}–$10^{-2}\%$ is a lethal dose for larvae, sea invertebrates, crustaceans, oysters, snails, shrimps and fishes. Only a few marine plants can resist toxicant concentrations up to 10^{-2}–$10^{-1}\%$ [162].

Soil contamination

Local contamination of soil is more diverse and prolonged than that of water or air. The main factor controlling this phenomenon is the high adsorption capacity of soil, as well as the physico-chemical characteristics of the contaminants and particularly their solubility and resistance in a natural environment. The binding of toxicants is accomplished by both the inorganic part and the organic matter of soils. Among the minerals, strong adsorbents are clays of which the toxicant adsorption ability generally decreases in the following order:

kaolin > bentonite > silt.

Besides adsorption to humus, the binding of toxicants by hydrogen bonds and covalent bonds often takes place; therefore, toxic compounds in soil are retained longer by organic matter. For instance, it has been shown that 30% of the pesticide amiben (3-amino-2,5-dichlorobenzoic acid) introduced into soil is bound to humus, but only 10% is adsorbed by clay [277].

Binding to humus proceeds via the polar functional groups of toxicants (hydroxyl: amine, carbonyl, carboxyl, etc.). These groups enhance the polarity of the toxicant molecules and enable the formation of hydrogen bonds between the toxicant and soil organic matter. They may also participate in the covalent binding of toxicants with humus components, e.g. with humic and fulvic acids. The release of soil-bound toxicants via elution or hydrolysis proceeds very slowly, and, therefore, soils remain contaminated for a long time.

Another reason of prolonged soil contamination is the stability of toxicants as determined by their chemical structure. The stability of aliphatic hydrocarbons is enhanced by saturation and branching of the chain. Aromatic hydrocarbons are more stable with respect to biotransformations, and the presence of substituted groups around the aromatic ring increases stability. Halogen-substituted aromatic hydrocarbons (especially when the substituents are chlorine and fluorine atoms) are the most stable.

The total removal of toxic compounds from the environment proceeds only by their full mineralization, i.e. when the organic substrates are decomposed to CO_2, H_2O, HCl, NH_3 and some other inorganic substances [162]. Such degradation of toxicants in soil can be accomplished via both abiotic and biotic pathways. Abiotic transformations include photochemical and chemical oxidation-reduction processes, as well as hydrolytic splitting. Soil organic matter, metal oxides and minerals participate in these processes. In the soil the following processes, *inter alia*, proceed abiotically: reducing dehydrochlorination of the insecticides lindane and DDT, reduction of nitro groups to amides in parathion and pentachloronitrobenzene, and saponification of organophosphorus insecticides. The main pathway of full destruction of toxic organic compounds is their biological mineralization, i.e., degradation by microorganisms capable of using these substances as a nutritional carbon source.

The rate of microbiological toxicant decomposition depends on exogenous factors such as oxygen concentration in the soil, temperature, soil pH, the presence of inorganic and organic nutrients, etc. Among these factors the content of oxygen in soil, which limits the intensity of anaerobic microorganism growth, is the most significant.

Stability, i.e. persistence of toxic compounds, is estimated by the time needed for transformation of 95% of the toxicant. Persistent organic contaminants (POPs) are: dioxins, PCBs, most organochlorine pesticides (aldrin, dieldrin, endrin, chlordane, lindane, heptachlor, hexachlorobenzene, mirex, toxaphene, DDT, etc.), PAHs etc. For dioxins the period of 95% destruction is 14–15 years, for PCBs it is 10–12 years, for DDT 4 years, for heptachlor 3.5 years, for lindane 3 years, etc. Widely distributed *sym*-triazine pesticides (simazine, atrazine, prometryn) are retained in soil for about 2 years, carbamates from several months to a year, and organophosphorus insecticides (chlorophos, metaphos, etc.) and derivatives of phenoxyacetic acids (2,4-D, 2,4,5-T, etc) are destroyed within several months [277].

In most cases the process of mineralization of POPs requires the joint action of anaerobic and aerobic microorganisms in a certain sequence. As a rule, first chlorine atoms have to be removed or substituted by hydroxyl radicals before further degradation can proceed. Anaerobic microorganisms in oxygen-poor soil generally accomplish such transformations, and, as a result the aromatic ring of the initial toxicant remains. This ring is accessible only for the oxidases of aerobic microorganisms (Mn-peroxidases, phenoloxidases, laccases, etc.) [257, 287].

Chemical accidents

Industrial accidents and catastrophes with highly toxic substances have become an evil of civilization. How serious and hazardous they are for the environment can be easily traced from examples of chemical accidents with chlorine, the first chemical weapon of the 20th century. Gaseous chlorine is 2.5-fold heavier than air. Some known accidents caused by chlorine in the former USSR are briefly described below [161]:

- Dzerzhinsk (Nizhegorod region): July 1947, a cistern flange junction broke, chlorine was dispersed by the wind, 28 men were poisoned (13 lost the ability to work); September 1967, a chlorine pipeline broke; January 1974, a defective pipeline resulted in the ejection of chlorine from a tank, 11 persons were heavily poisoned; March 1986, leakage of liquid chlorine from a product valve.
- Volgograd: February 1966, group poisoning with chlorine, 8 victims; June 1966, group poisoning with chlorine, 48 victims, 7 hospitalized; December 1966, ejection of 1.9 tons of chlorine from a tanker, 115 victims within a radius of 1500 m, 22 hospitalized; March 1968, accident resulted in heavy poisoning with chlorine, 5 victims; August 1969, heavy poisoning with chlorine as a result of a pipeline break in a fire; 8 victims, 4 with loss of the ability to work; January 1985, leakage of 7.5 tons of chlorine from tankers, 27 victims; May-June 1988, spilling of chlorine from two wrecked tankers.
- Berezniki (Perm region): August 1976, 38 tons of liquid chlorine spilled from an open valve, 27 victims within a radius of 1 km, 2 died.
- Novomoskovsk (Tula region): January 1987, leakage of liquid chlorine, 1 person died; June 1987, chlorine leakage from two wrecked tankers; July 1987, release of 20 tons of liquid chlorine; 1989, chlorine leakage from an overfilled tanker.
- Usole-Sibirskoye (Irkutsk region): December 1955, accident at a chlorine pipeline, 53 victims, 17 hospitalized; March 1964, chlorine leakage from a pipeline, chlorine cloud moved toward the neighboring town and covered a residential area of 3 x 4 km, 75 hospitalized.
- Zima (Irkutsk region): December 1987, release of liquid chlorine; 2 poisoned and burned with chlorine; June 1981, leakage of liquid chlorine from a tanker; 1986–1987, 11 accidents around a train with chlorine cisterns.
- Kuibyshev, (Novosibirsk region) May 1990, chlorine leakage from storage tank at a chemical plant.
- Sterlitamak (Bashkiria): December 1968, heavy group poisoning with chlorine as a result of an accident, 60 victims, 46 hospitalized; December 1989, chlorine leakage from over-filled tankers.

– Zaporozhye (Ukraine): January 1989, release of 6 tons of chlorine, 1 victim.
– Rubeznoe (Ukraine): July 1971, release of 35 tons of chlorine.
– Khabarovskoe: June 1989, leakage of 800 kg of chlorine, 7 persons hospitalized.

Analogous accounts of inadvertent releases of phosgene, hydrocyanic acid, lewisite and other extremely hazardous toxic compounds can be found. The probability of industrial accidents and catastrophes with toxic agents was typical for Dzerzhinsk, where chemical weapons were produced.

Among the greatest recent chemical catastrophes in the world are: 1976, in the chemical plant of Seveso (Italy), where a violation of safety procedures resulted in the pollution of a territory over 18 km^2 with dioxin; about 1,000 people suffered, and mass death of animals occurred. The remediation of the direct consequences of the accident lasted more than a year; one of the worst accidents ever in a chemical plant took place at Bhopal (India) in 1984: 3,150 persons died and over 200,000 were mutilated to the first degree. The cause was the release of methyl isocyanate; in 1988 three thousand people suffered from spilled toxic agents in a railway accident in Jaroslavl (USSR); in 1989, a chemical accident occurred in Ionave (Lithuania): seven thousand tons of liquid ammonia spilled on the territory of a plant, forming a lake of toxic liquid with an area of about 10,000 m^2. A fire started and burnt down a store with inorganic fertilizers including nitrates, whose thermal decomposition was accompanied by the emission of toxic gases. The air pollution reached a height of 30 km but due to favorable meteorological conditions the cloud of pollution moved over unpopulated regions and did not harm many people; in August 1991 in Mexico 32 tankers filled with liquid chlorine derailed and about 300 tons of chlorine was released into the atmosphere, affecting about 500 men, among whom 17 died.

Despite increasingly stringent legislation directed to ensure safer working conditions, the global trend for such accidents is increasing. Protective measures are always running behind manufacturing practice, and there are good reasons to expect the occurrence of accidents with highly deleterious consequences to continue to increase in frequency. Factors contributing to this situation are:
– The unpredictable growth of industries using novel, more advanced technologies, based on high concentrations of energy and harmful substances.
– The enormous accumulation of different types of industrial wastes that are harmful for the environment.

– The absence or insufficient level of preventive measures.
– The inevitable increase in chemical industry, and the concomitant transportation and storage of harmful substances.
– Investment in harmful industries in the developing and consequently technically weakly developed countries, unable to implement proper environmental protection measures.
– Terrorist attacks on chemical plants.
– Risk assessment and minimization (RAM) of technical systems are mainly carried out using traditional rules of forecasting. There is a great need for a decisive paradigm shift in RAM.

The above-presented data concern only one harmful category of human activity – accidents in chemical industries. The ecological harm engendered by military activities practically in all countries of the world must also be emphasized. These activities usually are accompanied by an increased pollution with toxic compounds at training and firing grounds, and in war zones.

Another very significant source of toxic compounds is road and air transportation, daily emitting large quantities of products resulting from incomplete oil combustion.

The constantly increasing level of toxic compounds has a negative effect on nature, especially on such vitally important biological processes as respiration, photosynthesis, fixation of molecular nitrogen, reproduction, etc. High concentrations of environmental contaminants may cause mutagenesis that lead to the destruction of particular living species and the creation of new ones, often degenerate and weakened.

2 Uptake, translocation and effects of contaminants in plants

2.1 Physiological aspects of absorption and translocation of contaminants in plants

Environmental contaminants enter the plant cells from air, soil and water. Plants absorb contaminants primarily through their roots and leaves. Contaminants enter leaves as a result of the direct spraying of plants with agrochemicals and by absorption of gaseous contaminants in the air. Below ground, xenobiotics penetrate plants together with water and nutrients through the roots. Chemical compounds are absorbed by roots less selectively than by leaves. The processes of toxic compound absorption by leaves and roots differ essentially from each other. Studies in this area were conducted especially intensively in the period 1965–1990.

2.1.1 Absorption of environmental contaminants by leaves

In order to penetrate into a leaf, the xenobiotic should pass through the stomata, or traverse the epidermis, which is covered by a wax cuticle (Fig. 2.1). Generally, stomata are located on the lower (abaxial) side of a leaf, and the cuticular layer is thicker on the upper (adaxial) side.

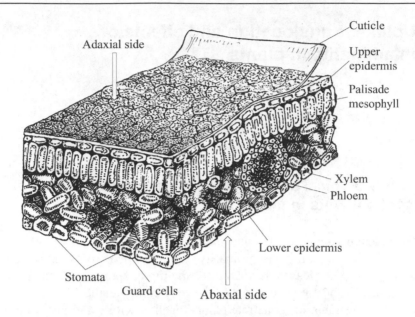

Fig. 2.1. Structure of the leaf

The stomatal system serves as a major regulator in the xenobiotic penetration process in leaves. By changing the aperture diameter the plant con-
trols the entry of compounds of different molecular masses. Opening and
closure of the stomata is controlled by movement of two special kidney-
shaped guard cells (Fig. 2.2), modified epidermal cells [310].

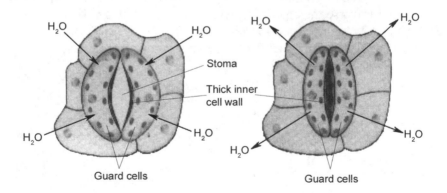

Fig. 2.2. Structure of stomata

The movement of these guard cells is regulated by the concentration of
potassium ions. The stoma is opened by increasing K^+ concentration. The

extent to which the stoma is opened depends on external environmental conditions, such as light, temperature, humidity etc., and on internal factors, such as the partial CO_2 pressure in the intracellular space, the plant hydration condition, the ionic balance and the presence of pheromones [310].

Gases and liquids penetrate through stomata into leaves. The permeability for gases depends on the extent to which the stomatal apertures are opened (4–10 nm), and the permeability for liquids depends on the moistening of the leaf surface, the surface tension of the liquid and the morphology of the stomata. The majority of toxic compounds penetrate into a leaf as solutions (pesticides, air pollutants, liquid aerosols etc.). For the leaves of zebrine (*Zebrina purpusii*) it was established that the crossover surface tension of their lower surface is 25–30 dyne/cm (for comparison: the surface pressure of water equals 72.5 dyne/cm, for ethanol 22 dyne/cm) [441]. Liquids with a surface tension less than 30 dyne/cm have a constant angle of contact with the surface of a leaf and instantly penetrate into the stomata. Liquids with a surface tension of >30 dyne/cm are able of penetrating into the stoma without moistening the leaf surface.

The penetration of α-naphthylacetic acid into the stomata–rich lower surface of pear (*Pyrus* sp.) leaves was considerably enhanced under the influence of light [200]. The growth regulator α-naphthylacetic acid and natural phytohormones significantly influence the mechanism of stoma opening. Experiments on isolated leaves of light blue snakeweed (*Stachytarpheta indica* (L.) Vahl) showed that α-naphthylacetic acid and its hydroxyderivative metabolite, 2-naphthoxyacetic acid, restrict, but do not prevent, the opening of stomata and suppress the accumulation of potassium in guard cells [370]. The penetration of succinic acid 2,2-dimethylhydrazide through the surface of isolated leaves of kidney bean (*Phaseolus vulgaris)* also indicates the relevant role of the stoma in the absorption of xenobiotics by leaves [442]. The process of absorption is stimulated by illumination.

However, according to some publications the role of stomata in the absorption of xenobiotics is rather insignificant. In addition, often contrasting data for the same compounds are published. For example, the presence of stomata on the leaf surface did not appear to influence the penetration of the herbicide 2,4-D [103]. On the other hand, the penetration of this herbicide into the leaf was increased by intensive illumination. Evidence for the penetration of herbicides through stomatal apertures is provided by the observation that absorption of 2,4-D is more intensive on the abaxial side of hypostomatic leaves than the adaxial [430]. Experiments with various ecotypes of creeping thistle (*Cirsium arvense*), differing in the quantity of stomata per unit leaf area, have shown that the penetration of 2,4-D into leaves does not depend on the surface area of stomatal apertures and the number of stomata [233].

For inorganic gases penetration through the stomata is the main pathway; for example, CO_2 is absorbed 100 times more intensively through stomata than through a cuticle. For plants without stomata, e.g. ferns, water plants and algae, penetration through the cuticle of the epidermis is the sole path for xenobiotics to enter the leaves.

The cuticle is a wax layer covering most of those parts of higher plants that protrude from the soil. The cuticle is permeable not only for lipophilic substances, but also for hydrophilic molecules of gas, water and sulfuric acid [310].

The thickness and chemical composition of the leaf cuticle vary with species, age, and location on the stem as well as with environmental factors such as temperature, humidity, etc. The cuticle consists largely of wax (cutin), which is a complex mixture of long-chain alkanes, alcohols, ketones, esters and carboxylic acids. Alkanes and esters predominate on the outside surface of a cuticle. Besides, long-chain C_{29}–C_{33} diketones, triterpenoids (for example, ursolic acid), diterpenes, glycerides and phenolic compounds are sometimes present in the wax. The main mass of leaf wax is attributable to normal long-chain alkanes with an odd number of carbon atoms in the chain (C_{31}–C_{37}), in particular the n-alkanes $C_{29}C_{60}$ and $C_{31}C_{64}$, and esters of n-carboxylic acids with primary and secondary alcohols [144].

The thickness of the cuticle varies and depends not only on the plant species, but also on the age of the leaf. Thickness is not always an indicator of wax content. For example, in two varieties of plum (*Prunus domestica*) the adaxial surface of leaves have a thicker cuticle (1.6–2.0 nm) than the abaxial one (1.2 nm). However, the thicker one contains less surface wax (densities 34–35 and 47–52 $\mu g/cm^2$, respectively) [303]. In young leaves the cuticle is usually thinner and less uniformly developed than in old ones. The quantity of stomatal wax also increases with ageing. The synthesis of cutin is terminated only after complete leaf greening [274]. This explains why young leaves absorb toxic compounds much more intensively than mature ones. For example, the absorption of succinic acid 2,2-dimethylhydrazide through the surface of kidney bean leaves decreases with leaf age [442]. Similar results have been reported for 2,4-D, nitrofen, indole acetic acid (IAA) and some other herbicides [153, 372, 430].

Environmental contaminants adsorbed on the lipophilic surface of leaf wax accumulate in the cuticle to a large extent and gradually penetrate into the leaf cell system. The wax appears to be an active sorbent for lipophilic toxic compounds [58]. Apparently, the molecules of the adsorbed xenobiotics together with individual wax components migrate from the cuticle inside the epidermal cells and are incorporated into endocellular membranes [68]. The wax layer of a cuticle serves as a barrier for the adsorption and penetration of organic compounds into leaves. This is seen in isogenic lines of pea (*Pisum sativum*) possessing or lacking a gene determining the formation of wax (phenotypically this difference is expressed via cuticle thickness). The pesticides carbophos and methylnitrophos rapidly penetrate into the leaves of waxless plant lines, reaching the photosynthesizing tissues in 3-4 h [359]. In plants with wax-coated leaves these pesticides penetrate only to a small extent. In another example, the removal of wax from the surface of apices of European furze (*Ulex europaeus*) seedlings causes a 3.5-fold increase in the absorption rate of ^{14}C-picloram (a systemic herbicide) [414].

The ease with which a contaminant penetrates a cuticle largely depends on its physical properties, above all on its lipophilicity. For instance, in contrast to the pesticide pyrazon, which promptly penetrates into the leaves of red beet (*Beta vulgaris*), phenmedipham and benzthiazuron with far lower lipophilicities are absorbed in insignificant amounts [336]. Many organic contaminants are capable of changing the composition and structure of cuticular wax by increasing the permeability of the cuticle. The herbicide bioxone is characterized by an extremely high rate of penetration through the cuticle [258]. The cuticle is also permeable to larger molecules, such as some surfactants, fatty acids, long-chain alkanes, peptides, salts of 2,4-D with long-chain amines, etc. [68, 157, 455]. However, the correlation between permeability and molecular mass of the contaminants is poor.

In the process of infiltration, the aggregation state of the toxic compound is very significant. According to data obtained with the fungicide pyracarbolid, enhancing the degree of dispersion increases the permeability in the following order [419]:

powder < suspension < emulsion < genuine solution.

The intensity of penetration of ionogenic substance solutions into leaves depends on the pH of the solution. An interesting regularity is observed with such toxic compounds as picloram, 2,4,5-T, succinic acid 2,2-dimethylhydrazide and α-naphthylacetic acid. Xenobiotic molecules, particularly weak acids, are predominantly absorbed in their nondissociated state, as in this nonionic state they have a lower polarity and easily overcome the lipophilic wax barrier of the cuticle [25, 414, 442, 466].

Very often organic substances (contaminants) penetrate into damaged tissues of plants much more easily than into intact plants. For example, the amino acid derivative pesticide N-lauroyl-L-valine penetrates into the leaves and fruiting bodies of cucumber (*Cucumis sativus*) to a low extent [455], but which is increased in damaged leaves and fruits.

Possible pathways of the penetration of lipophilic organic contaminant in leaves have been clearly shown, viz. the absorption of gaseous hydrocarbons by hypostomatic leaves (with stomata only on the lower leaf surface) [508]. The leaves of the field maple (*Acer campestre*), wild Caucasian pear (*Pyrus caucasica*), vine (*Vitis vinifera*) and narrow-leaved oleaster (*Elaeagnus angustifolia*) were placed in an atmosphere containing ^{14}C-methane or [1–6-^{14}C] benzene. Contact with labeled hydrocarbon occurred only from one side of the leaf. The total radioactivity of the non-volatile metabolites formed showed that the absorption of gaseous alkanes and vapors of aromatic hydrocarbons by leaves occurred not only through the stomata, but also through the cuticle. The pathway through the stomata was preferred (Fig. 2.3).

Similar results have been obtained for a number of herbicides (α-naphthylacetic acid, 2,4-D, picloram and derivatives of urea), applied in soluble form to leaves [303, 430, 453]. The abaxial side of a leaf, rich in stomata, absorbed the organic substances more rapidly than by the adaxial side, possibly due to the thinner cuticle on the abaxial side. These results imply the active participation of stomata in the absorption of toxic compounds from ambient air.

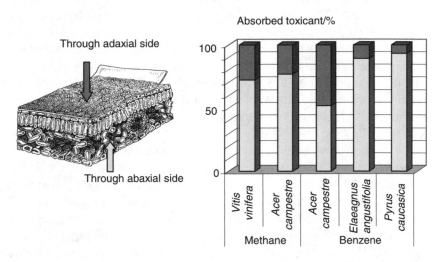

Fig. 2.3. Absorption of ^{14}C-methane (specific radioactivity 1 μCi/ml) and [1–6-^{14}C] benzene (specific radioactivity 4.9 μCi/ml) by the hypostomatic leaves of plants. The concentration of methane in the air was 1.5% by volume. Conditions: 8 h exposure under illumination, the concentration of benzene in the air was 2 mg/l, 4 h exposure in darkness [508]. Dark part of columns – through the adaxial side; Light part – through the abaxial side.

Trichomal cells (trichomes are outgrowths of the epidermis, such as filaments, warts, scales, setae etc.) can also participate in the absorption of toxic compounds. The absorption of the herbicide ^{14}C-triclopyr increased with the number of radial trichomes on the adaxial surfaces of young leaves of tanoak (*Lithocarpus densiflorus*) [270].

One of the probable mechanisms of toxic compounds entering in leaves is the penetration of xenobiotics through the ectodesmata, which are hollow tubes in cell walls, constituted from cellulose fibers. These canals connect the plasmalemma to a cuticle and can serve as conductive pathways both during the absorption of water-soluble substances by a leaf and during their excretion. For example, it has been shown that the adsorption of the herbicide 2,4-D (tritium labeled) takes place on the anticlinar walls of the epidermal cells, mainly at places where ectodesmata are located, and the absorption of this herbicide by the leaves of wheat (*Triticum vulgare*) and kidney bean is directly proportional to the number of ectodesmata per unit of epidermis surface area [173].

2.1.2 Penetration of contaminants into roots

The process of toxic compound absorption by roots differs essentially from the penetration of environmental contaminants into leaves. Substances pass into roots only through cuticle-free unsuberized cell walls. Therefore, roots absorb substances far less selectively than leaves.

Environmental contaminants enter the roots together with water, like nutrients. They move towards the transport tissue (xylem) mainly along the apoplast, a free intercellular space directed towards the xylem (Fig. 2.4). A comparatively small amount of contaminants moves along the symplast, through cells and the plasmodesmata that bridge cells.

Substances penetrate into the apoplast, a system of microcapillaries, by diffusion. They easily move through these capillaries and do not meet membrane barriers on their way, in contrast to symplastic transportation.

Roots absorb environmental contaminants in two phases [278]: in the first fast phase, substances diffuse from the surrounding medium into the root; in the second they slowly accumulate in the tissue. The intensity of the absorption process depends on the molecular mass of the contaminant, concentration, polarity, pH, temperature, soil humidity and some other factors [282, 514].

In this initial stage of absorption of environmental contaminants, the diffusive penetration of substances into the roots apoplast takes place. The rate of this process is directly proportional to contaminant concentration in the soil or nutrient solution.

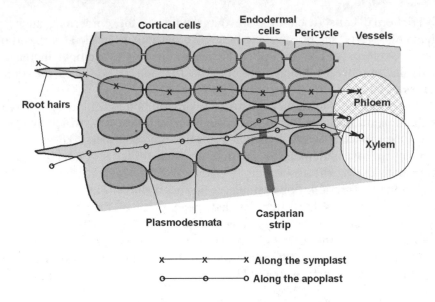

Fig. 2.4. Pathways of environmental contaminant penetration into roots.
In the transport of compounds along the apoplast, the water-impermeable
Casparian strip needs to be crossed and osmotic or symplastic routes must
be used.

Temperature greatly affects the absorption of organic contaminants by
roots. The temperature coefficient for diffusion processes (indicating how
much the reaction rate increases for a 10 °C temperature rise) is relatively
low (1.2–1.4). Therefore, the diffusion process practically does not depend
on temperature over plant-physiologically relevant temperature ranges.
However, passive diffusive absorption is followed by active transport, controlled by transpiration, metabolic responses and accumulation. A rise in
temperature inceases the transpiration rate and enzymatic reaction rates
(temperature coefficient 1.3–5.0), resulting in an increase in toxic compound absorption [278].

The molecular mass of a substance is the main limiting factor during the
passage of toxic substances into roots. Plants easily absorb organic substances with molecular masses ≤1000 [471]. Studies on polyethylene glycol absorption by the roots of cotton (*Gossypium hirsutum*) [296] and pepper (*Capsicum annuum*) [252] have shown that larger molecules also
penetrate into the roots. For example, small amounts of polyethylene glycol with molecular mass between 4,000 and 20,000 were also found in
plants. The amount of polyethylene glycol entering the plant was inversely
proportional to the polymer molecular mass. Polyethylene glycol enters

plants much faster and in significantly greater amounts if the roots are damaged. Polyethylene glycol absorbed by kidney bean and cotton seedlings is translocated through the plant without changing its molecular characteristics [9]. According to other data, roots can absorb high-molecular-mass compounds only after partial degradation of the molecules, however [181].

The majority of the high-molecular-mass humic acids (^{14}C-labeled) are adsorbed on the surface of roots and partly penetrate into the cells of the epidermis. This has been demonstrated in sunflower (*Helianthus annuus*), wild radish (*Raphanus sativus*) and wild carrot (*Daucus carota*). Smaller molecules of fulvic acids penetrate more deeply and reach the central cylinder of the xylem. However, the labeled carbon of fulvic acids does not penetrate into the overground parts of plants. Experiments with polyurethane have shown that the labeled carbon polymer is absorbed (after preliminary partial degradation of the polyurethane molecules in the soil) by tomato (*Lycopersicum esculentum*), cucumber, and strawberry (*Fragaria vesca*) root systems [180].

The roots absorb a rather wide spectrum of hydrophilic and lipophilic organic molecules (aliphatic and aromatic hydrocarbons, alcohols, phenols, amines, etc.). Even substances with a very low solubility in water, such as polycyclic hydrocarbons, are absorbed by roots [110, 121]. The influence of lipophilicity on the absorption process by roots is rather significant. Substances with a moderate hydrophilicity, i.e., with a $\log K_{OW}$ of 0.5–3.0, are more actively absorbed [277].

The pH of the soil or nutrient solution is a significant factor during the absorption of toxic compounds by the root system. In particular, the following factors controlling absorption by roots depend on pH:
 – Adsorption of contaminants by soil particles.
 – Mobility of contaminant molecules in the soil.
 – Degree of dissociation of ionogenic molecules.
 – Permeability of absorptive root tissues.

The absorption of the insecticide picloram by the roots of oat (*Avena sativa*) and soybean (*Glycine max*) serves as an example of the influence of pH on the entrance of xenobiotics into plant roots [250]. The amount of absorbed xenobiotic sharply decreases when the medium pH is changed from 3.5 to 4.5, whereas a further change in the range of 4.5–9.5 barely influences the absorption process. At pH 3.5 only 20% of the picloram is ionized, but at pH 4.5 the degree of ionization reaches 71.5%. Therefore, plant seedlings absorb the insecticide predominantly in nonionic form. Many toxic compounds in both leaves and roots are predominantly assimilated as undissociated molecules, i.e., without electrostatic charge [514].

Contaminant desorption from soil particles and its translocation in soil largely depends on pH. For example, the herbicide atrazine is better extracted from alkaline soil (pH 8.3), but other herbicides, such as chloramben and dicamba, are much better extracted under weakly acidic conditions (pH 4.1) [295].

Since the organic pollutants enter the roots together with water, their absorption rate by the root system is also determined by soil humidity. The toxic compound absorbed by plants decrease with soil water potential (see Chap. 2.1.4). Soil moisture plays an important role in contaminant absorption and desorption by soil. During the extraction of symmetric triazines, atrazine and chloramben from soil by organic solvents, the addition of water to the solvent considerably enhances the amount of the extracted herbicides and is a necessary condition for the complete extraction of contaminants [295].

Soil organic matter (e.g., humic and fulvic acids) also influences the absorption of organic contaminants by roots. The toxicity of the herbicides prometryn, fluometuron and trifluralin decreases with the increase in organic matter content in soil [522]. Similar results were found while the study of thirteen representatives of symmetric triazines (chloro, methylthio- and methoxytriazine) with oat seedlings [393]. The toxicity of hydrophilic contaminants decreased to a lesser degree than of lipophilic ones. The higher organic matter content may promote the adsorption of lipophilic herbicide molecules to soil, and, therefore, the process of contaminant absorption by roots is diminished and toxic concentrations in the plant cells are not reached.

The processes of transpiration and metabolism as well as mineral nutrition of plants are significant factors influencing the process of toxic compounds absorption by roots. For example, urea, a stimulator of transpiration enhances the absorption of atrazine by tomato roots [340]. Inhibitors of respiration, such as cyanide at a concentration of 10^{-3} M, reduce the absorption of trichloroacetic acid in wheat and oat seedlings by approximately 30%. 2,4-D, acting as an inhibitor of metabolism at a concentration of 10^{-3} M, suppresses the absorption of trichloroacetic acid by 70% and 54% in wheat and oat seedlings respectively [82]. The presence or absence of nutritional elements differently influences the uptake of toxic compounds into roots. For example, the uptake of methyl-2-benzimidazole carbamate and thiophanate-methyl into kidney bean seedlings is reduced by N, S and B deficiency in the nutrient solution. However, when deficient of N, B or K the uptake of parathion increases in the same plant [4]. The absorption of the herbicide buturon by wheat roots is decreased in N, P, K or Mg-deficient plants [216]. Mineral salts do not influence the uptake of benzimidazole and thiophanate (fungicides) in isolated maize (*Zea mays*) roots [305].

2.1.3 Penetration of contaminants into the seeds

The rate of pesticide penetration in seeds is limited by the permeability of the seed coat (Fig. 2.5). The amount absorbed also depends on other parameters, such as: concentration, temperature, duration of incubation, etc.

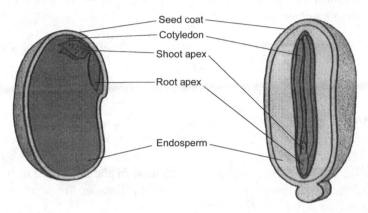

Fig. 2.5. Seed structure of dicotyledonous (left) and monocotyledonous (right) plants

For many species the seed coat is the main barrier impeding the penetration of xenobiotics into the internal seed organs. For example, cotton has a more impermeable seed coat than kidney bean, and accumulates the herbicides terbutryn and fluometuron mainly in the seed coat, while kidney bean allows both herbicides to penetrate the cotyledon of its seeds [416]. Dichloromethane has different effects on seeds of redroot pigweed (*Amaranthus retroflexus*) and oats depending on the duration of treatment and the physiological state of the seeds. Removal of the seed coat results in a decrease in the germination and respiration intensity in seeds [43]. Dichloromethane may not be capable of penetrating the seed coat. Studies on the influence of the lowest alkanes, alkenes, alkadienes and aldehydes on seed germination in weeds of purslane (*Portulaca oleracea*) and *Amaranthus* have shown that unsaturated hydrocarbons (ethylene, propylene and propadiene) easily penetrate seeds and stimulate their germination [495]. Aldehydes weakly stimulate and alkanes have no physiological effect.

In a study on the absorption of herbicides by the seeds of corn, soy and castor bean (*Ricinus communis*) from aqueous solution it was demonstrated that the seeds absorbed these compounds in the following order [447, 448]:

fluometuron < 2,4-D < diphenamid < trifluralin < atrazine < prometryn < chlorpropham.

It was hypothesized that absorption of herbicides in seeds is described by the following equation [447]:

$$M(t) = \frac{4\pi a^2}{W_s}(C_0 - C_1)\left[\frac{Dt}{a} + 2\left(\frac{Dt}{\pi}\right)^{1/2}\right] \quad (1.3)$$

where $M(t)$ is the amount of herbicide (in µg/g) absorbed by a seed; t the incubation period (in s); a the seed radius (in cm); W_s the air-dry weight of a seed (in g); D the diffusion coefficient of the herbicide in the absorption medium (in cm^2/s); C_0 the initial concentration of the herbicide in the medium (µg /ml), C_1 the concentration of the herbicide at the seed coat surface (µg /ml).

A study out the absorption of various ^{14}C-labeled herbicides – amiben, atrazine, monuron, S-ethyldipropylthiocarbamate and chlorpropham – in soybean seeds has shown that there is a direct relation between the herbicide concentration and its absorption [404]. Absorption increases with increasing temperature between 10 and 30 °C. The absorption was almost identical in live and dead seeds.

From these examples we may conclude that the penetration of organic contaminants of different chemical structures into plant cells follows a regular pattern. Compounds that penetrate the cell become substrates for enzymes able to transform them. The abilities of plants to permit contaminants of a certain structure, lipophilicity, molecular mass and charge, to enter their intercellular space are the initial characteristics determining the detoxification potential of plants.

2.1.4 Translocation of environmental contaminants in plants

Once absorbed by roots and leaves, environmental contaminants are translocated to different plant organs by the same physiological processes transporting nutrients. The main processes involved are:

− Transpiration: transport of water and dissolved substances from roots to shoots, passing through vessels and tracheids in the xylem.
− Assimilate transport: transport of assimilates and dissolved substances from the leaves to the parts of a plant below (shoot axes, roots) and above (shoot tops, fruits) the leaves, passing through sieve tubes in the phloem.

The transpiration streaming is engendered by the evaporation of water by leaves. Transpiration or evaporation of water by the plant is a diffusion process driven by the kinetic energy of water molecules and the gradient of water potential in the plant/air system. All external factors raising this gradient enhance the transpiration process. For instance, an increase in temperature results in a decrease in the relative humidity of the air, causing an increase in transpiration intensity. The wind promotes the opening of stomata and enhances transpiration by quickly removing water from the leaf surface by evaporation.

The main physical force driving the transpiration stream is the large difference between the water potentials of atmospheric air ($\Psi_w \sim -900$ atm at 50% humidity) and soil (Ψ_w is usually close to zero). The water potential (Ψ_w) is always negative and represents the difference between the chemical potential of water at a definite point (Ψ_w) and pure water (Ψ_{ow}), divided by the partial molar volume of water ($\overline{V}_w$) [310]:

$$\Psi_w = \frac{\mu_a - \mu_w}{V_w} \qquad (1.4)$$

The water potential of a plant is an average between that of the air and soil that determines the leakage of water from soil to the atmosphere through the plant via evapotranspiration.

Transpiration protects the plant from overheating. Both stomata and cuticles participate in this process. The majority of the water is evaporated via the stomata into the atmosphere. Cuticular transpiration typically comprises not more than 10% (in plants with a thin cuticle it may reach 20%). The total area of the stomatal apertures constitutes only 1–2% of the leaf surface, but even this comparatively small area enables a significant amount of water to be evaporated. For instance, an average birch tree (*Betula*) evaporates approximately 400 l of water per day, a poplar from 190 to 1330 l, a willow (*Salix alba*) 1900 l, the latter being about the same amount as that evaporated from alfalfa occupying 0.243 ha [188].

The mechanism of transpiration ultimately driven by the evaporation of water from the leaves is schematized in Fig. 2.6. Mesophyllic cells, located in the leaf and in contact with the air via the stomatal aperture, lose water in the process of transpiration. As a result, there is a water deficiency in the cell walls, which is very rapidly distributed over the whole cell. The loss of water induces compression of the cell wall, a decrease in the volume of the cytoplasm and shrinkage of the vacuole. Consequently, the Ψ_w of the cells increases and the osmotic pressure rises. In other words, cells located near stomata develop a suction force caused by the difference of water potential between cell and atmosphere that takes away water from the neighboring cells, which in turn develop a suction force. Thus, the suction force reaches the xylem, passing through the leaf vessels. The xylem constitutes a permanent tube system connecting leaves to roots; as a result, the suction force in xylem drives the water penetration from roots to xylem.

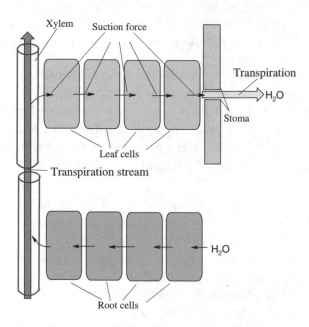

Fig. 2.6. Diagram illustrating the direction of the transpiration stream in plants. Note: the phloem system, i.e. vessels parallel to the xylem, does not participate in the upward flow.

The suction force of transpiration pulls the water column up, filling the vessels at rates from 1 to 100 m/h. The rate depends on the plant species, height and environmental conditions. During the movement of water through narrow vessels, a pressure of about 1.5 atm per 10 m height is needed to overcome the flow resistance. Transpiration occurs even in trees, with a height of >100 m. For water to rise to such a height a cohesive force between water molecules is required to prevent them from entering the gas phase. A pressure of 35 atm would enable a cohesive water column of about 140 m, and, consequently, the height of the tallest trees does not exceed this height [310].

The daily transpiration stream depends on temperature: the highest flux occurs soon after noon and the lowest at night. This process is accompanied by the opening and closing of the stomata, respectively, in the light during the day and in darkness at night. During hot, dry days the evaporation rate of water from the leaf surface exceeds the flow of water from the roots. The resulting water deficiency leads to deformation of the leaves, which tends to narrow the apertures of the stomata, leading to a decrease in the transpiration rate. At nightfall, the stomata are closed due to the absence of light, the ambient temperature falls, the water evaporation rate decreases, and water deficiency in the roots does not occur.

Many morphological peculiarities serve to minimize the transpiration rate, e.g. wax on fruits and leaves, a dense cuticle, a suberic layer, stomata submerged in the mesophyllic layer, and reduction in leaf surface area. All these features are characteristic of plants growing in dry areas (xerophytes). Plants growing in humid areas (hydrophytes) have mechanisms for intensifying transpiration, i.e. a large leaf surface area, elevated stomata, a thin cuticle, etc. Submerged and floating plants (hydrophytes) may absorb water across their entire submerged surface area. They have a primitive xylem vessel system and the water flow rate is insignificant.

The processes described above are attributable to higher plants capable of maintaining their own humidity (homeohydrophytes). The opposite behavior is found in lower, and in some less widely distributed higher plants with variable cell humidity (poikiloxerophytes). In the latter plants the water potential is in balance with the environment, the absorption of water is inseparable from its excretion and almost no net water flow occurs.

The importance of the transpiration stream for the absorption and translocation of environmental contaminants in plants is illustrated by the equation proposed for calculating the rate of contaminant assimilation from polluted soils [437]:

$$U = (TSCF)\,(T)\,(C) \qquad (1.5)$$

U is the rate of contaminant assimilation (mg/day); T the rate of plant transpiration, (l/day); C the contaminant concentration in the water phase of soil (mg/l); $TSCF$ the transpiration stream concentration factor, dimensionless, showing the ratio between the organic contaminant concentration in the liquid of the transpiration stream and its concentration in the environment [368]. The $TSCF$ depends on the physical and chemical characteristics of the contaminant and can be estimated by the empirical equation [61]:

$$(TSCF) = 0.75\exp\left[-\frac{(\log K_{ow} - 2.5)^2}{2.4}\right] \qquad (1.6)$$

From equation 1.6, the main parameter characterizing the contaminant is K_{ow}, the partition coefficient between octanol and water. This parameter gives an indication of the hydrophobicity that predetermines the effectiveness of absorption and translocation of a contaminant in plants. It is known that contaminants with a $\log K_{ow} > 3.5$ are well adsorbed on soil granules or plant root surfaces and do not penetrate into the plant interior. Examples of such environmental contaminants are 1,2,4-trichlorobenzene, 1,2,3,4,5-pentachlorophenol, PAHs, PCBs, dioxins, etc. Moderately hydrophobic contaminants with a $\log K_{ow}$ between 1 and 3.5 (phenol, nitrobenzene, benzene, toluene, TCE, atrazine, etc.) are absorbed in large quantities and penetrate more easily into the plant. Hydrophilic contaminants with a $\log K_{ow} < 1$ (aniline, hexahydro-1,3,5-trinitro-1,3,5-triazine (the explosive RDX), etc.) are slightly adsorbed and not intensively assimilated by plants [437].

An environmental contaminant translocates along the plant via the transpiration stream after penetration. However, xenobiotics absorbed by the leaves are translocated together with the stream of assimilates formed in leaves. As indicated above, the transpiration stream passes through the xylem, and the stream of assimilates through the phloem. The xylem is a unidirectional flow vessel from roots to shoots, the phloem being a bidirectional flow vessel. Toxic compounds absorbed by roots can migrate to the xylem through the apoplast in the following way:

roots hairs → intracellular spaces → cell walls of cortical cells → endodermis → passing the Casparian strip barrier → xylem.

The Casparian strip (see Fig. 2.4) is located around the central channel of the vessels, and creates additional resistance for the transportation of toxic compounds. The Casparian strip has the same function in roots as the waxy cuticular layer in leaves – it provides protection from water deficiency.

The sieve tubes of the phloem consist of special cells with undeveloped vacuoles connected with each other by special pores, and connected with the symplast by the plasmodesmata. In these cells, the cytoplasmic pH varies from 7.0 to 7.5. The pH of phloem juice is about 8.0 and that of vacuoles is about 5.5. The pH of the apoplast, including the intracellular space and the xylem vessels, is in the range of 5.0–6.0. In the xylem, the water streams from 50 to 100 times faster than in the phloem. Streaming velocity in the phloem varies from 0.5 to 1.0 m/h [45].

Assimilates pass through the phloem that consists of sieve tube segments composed by adjoined cells. The lateral walls between the segments are called sieve plates (Fig. 2.7). The sieve plates are pierced by widened plasmodesmata (transport pores). Cells of sieve plates are peculiar: their nuclei are degenerate, plastids undeveloped, and mitochondria small and few. These cells contain no tonoplasts, and vacuole and protoplasm are not separated from each other. Phloem cells have a central lumen similar to a vacuole, filled with loose protoplasm, surrounded by a peripheral layer of dense protoplasm.

The transport of assimilates requires energy. This process is suppressed by oxygen deficiency, by a decrease in temperature and by disconnectors or inhibitors of respiration. Assimilates (glucose, amino acids, etc.) reach the so-called cell satellites and surrounding sieve tubes of parenchymal cells from the place of their formation via the symplastic pathway. Initially the synthesis of sucrose takes place in these parenchymal cells, rich in phosphatase, followed by the transport of assimilates into the sieve tubes through active transport. At the sites of consumption or deposition (storage), assimilates pass through the sieve tubes also by means of active transport and reach the individual cells via the symplast.

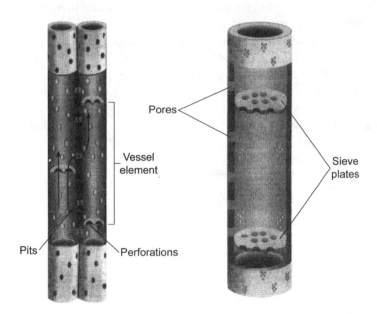

Fig. 2.7. Elements of xylem (left) and phloem (right) vessels

The stream of assimilates originates as follows: high osmotic concentrations in the places of assimilate formation enhance the osmotic absorption of water and induce a high hydrostatic pressure. Simultaneously, low osmotic concentration, insignificant osmotic absorption of water and a weak hydrostatic pressure are observed in the neighboring cells when assimilates are excreted. Hence, the movement of solution in the direction of the concentration gradient through the semipermeable membrane and between the sieve tubes and the surrounding cells equalizes the pressure.

Figure 2.8 summarizes the pathways of penetration and translocation of toxic compounds in plants.

Environmental contaminants in soil can bind to soil particles and this can be reversible or irreversible, depending on the hydrophobicity of the compound (the value of K_{ow}). Part of the contaminant can undergo microbiological transformations by rhizospheric microorganisms, the other part can penetrate the roots and migrate into the xylem vessels. Metabolites and intermediates produced from contaminant transformation by microorganisms can also penetrate into roots. Once absorbed, the xenobiotic is translocated by the transpiration stream and distributed throughout the plant. Contaminants that penetrate into the leaves through stomata or the cuticle or both reach the sieve tubes of phloem and translocate together with assimilates basipetally to the roots, or acropetally to shoot tips and fruit, or to both directions. A substantial body of experimental data confirms the translocation of toxic compounds in plants. Plants exposed to low concentrations of C_1–C_5 alkanes, cyclohexane, benzene, and toluene absorb these substances and effect their deep oxidation. In experiments involving the incubation of different plant species (55 annual and perennial plants) with labeled (^{14}C) hydrocarbons, it has been shown that all tested plants absorb and transform alkanes and aromatic compounds, but with different efficiencies [131, 133]. The products of hydrocarbon transformation assimilated initially by the leaves, move along the stem to the roots, but hydrocarbons absorbed and transformed by the roots are transported to the leaves (Fig. 2.9 and Fig. 2.10) [511].

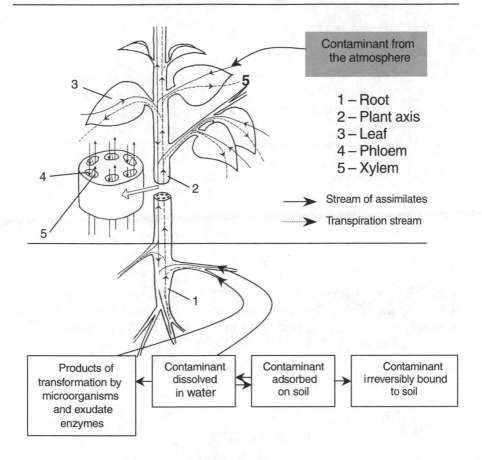

Fig. 2.8. Environmental contaminant penetration and translocation pathways in plants

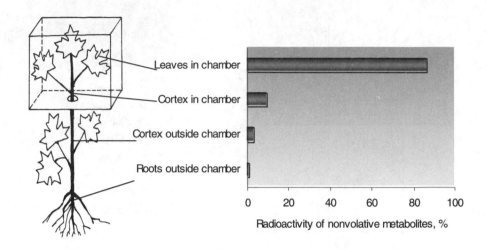

Fig. 2.9. Translocation of the radioactive atom of [1–6-^{14}C] benzene, absorbed by the leaves of maple (*Acer campestre*). Specific radioactivity of benzene, 5.5 µCi/mg; concentration in air, 2 mg/ml; temperature, 21 °C; duration of exposure, 96 h; in the dark [511]

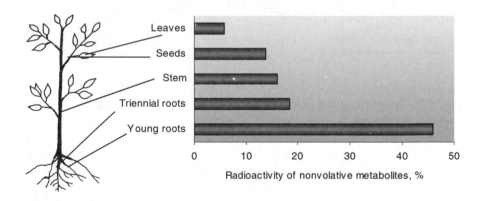

Fig. 2.10. Translocation of the radioactive atom of [1–6-^{14}C] benzene absorbed by the roots of tea (*Thea sinensis*). Specific radioactivity of benzene, 2.5 µCi/mg; saturated solution in water; temperature, 20–30 °C; duration of exposure 72 h; in the dark [511]

Many PAHs are actively absorbed and transported by both roots and leaves from the nutrient solutions despite their high hydrophobicity [108,

111, 234, 349]. PAHs that contain two or three rings and have a low molecular mass (e.g., naphthalene, anthracene, phenanthrene) are absorbed and degraded more easily than PAHs with a high molecular mass (e.g., perylene, 3,4-benzopyrene, benzanthracene, dibenzanthracene).

Aryloxycarboxyl acid pesticides penetrate the leaf cuticle in the form of undissociated molecules and are absorbed by the parenchymal cells. These xenobiotics reach the phloem via the symplast and enter the leaves, actively growing tissues, and reproductive organs via the sieve tubes. The herbicide 2,4-D and defoliant 2,4,5-T are translocated basipetally and acropetally to the rest of the plant after entering into the (kidney bean) leaf [317]. The herbicide mecoprop flows from leaves to roots with equal intensity in sensitive and resistant biotypes of the common chickweed (*Stellaria media*) [93].

Pesticides derived from urea are easily absorbed from nutrient solution by plant roots and most of them are quickly translocated upwards into the plant by entrainment in the transpiration stream. Cotton and kidney bean seedlings absorb ^{14}C-fluometuron from nutrient solution through the roots, and the transpiration stream quickly translocates the xenobiotic acropetally. This herbicide, when absorbed by leaves, moves basipetally and acropetally; symplastic translocation herbicide along the phloem is indicated [416]. Another derivative of urea, the herbicide tebuthiuron, is characterized by a similar migration route [479]. Chlorimuron, applied to the leaves of soybean, peanut (*Arachis hypogaea*) and various weeds, displayed a weak symplastic and apoplastic migration [530]. It is interesting to note that this herbicide, absorbed by the root system of yellow (*Cyperus esculentus*) and purple nutsedge (*Cyperus rotundus*) flows to the plant shoots, but if absorbed by the tubers is not translocated [398].

For carbamate-based pesticides acropetal translocation is typical. Examples are carbofuran in seedlings of soybean and mung bean (*Vigna radiata*) [491], and methyl-2-benzimidazole carbamate in seedlings of peanut [386, 519] and safflower (*Carthamus tinctorius*) [328]. The pesticides phenmedipham and desmedipham are translocated only acropetally after penetrating the leaves of wild mustard (*Brassica kaber*), *Amaranthus*, and sugar beet (*Beta vulgaris*), [228].

The direction of translocation may depend on plant resistance to the toxicant. For instance, the herbicide buthidazole, absorbed by leaves of sensitive *Amaranthus*, is translocated both acropetally and basipetally, but in resistant maize leaves transport proceeds only basipetally [223]. This herbicide is insignificantly translocated along the apoplast in soybean leaves [209]. 4,4′-Methylene-*bis*(2-chloroaniline) is absorbed but not translocated after application to the leaves of various plants.

Tyree et al. [507] suggested that xenobiotics could be transported in the phloem, based on the hypothesis of intermediate permeability. This hypothesis takes the immediate proximity of the phloem and xylem vessels into account and proposes that:

- Any molecule with a high permeability through the membrane will be able to get into the phloem, but also can leave the phloem and be more rapidly transported to the xylem stream.
- Any molecule with a low permeability through the membrane can not reach a sufficiently high concentration in the phloem for effective transport.
- There must be an intermediate permeability between these extreme values, and substances with such permeability should be characterized by the highest phloem mobility.

On the basis of this hypothesis, and a broad range of experimental data obtained from investigations of herbicide assimilation and transport by seedlings of castor bean, Kleier and coworkers have put forward a mathematical model enabling the determination of translocation of toxic compounds in plants [49, 199, 238, 239, 271]. Kleier's model has been successfully used in the prediction of translocation of many secondary metabolites: gibberelline A [364], salicylic acid [538], oligogalacturonides [405] and glucosinolates [50].

Systemic herbicides, i.e., herbicides capable of penetrating throughout the plant, based on their ability to easily move along the plant transport pathways, are divided into: phloem-mobile, xylem-mobile, and ambimobile. Ambimobile means capable of penetrating into both, phloem and xylem. Assignment to one of the above-mentioned classes is based on physico-chemical parameters such as the dissociation constant (pK_a) and lipophilicity (K_{ow}). Herbicides with a degree of dissociation characteristic of strong and medium acids ($pK_a < 4$) and with a medium lipophilicity ($\log K_{ow}$ about 1 to 2.5–3) belong to the phloem-mobile type, while weaker acids with $pK_a > 5$ and non-ionized compounds must be more polar to move well. Ability to translocate only along the xylem is characteristic for herbicides with a medium lipophilicity ($\log K_{ow}$ in the range of 0–4) and with a low degree of ionization ($pK_a > 7$) and a higher lipophilicity corresponds to a higher degree of ionization. Weak acids ($pK_a > 7$) with high hydrophilicity ($\log K_{ow} < 0$) are ambimobile, and highly lipophilic herbicides ($\log K_{ow} > 4$), regardless of the value of pK_a, are often non-systemic (so-called contact herbicides), because they can not translocate into the xylem or phloem [45].

Herbicide translocation in the phloem depends on the processes of carbohydrate biosynthesis (predominantly sucrose) in tissues and streaming from mesophyllic cells to the phloem cell system. In higher plants at least two mechanisms of carbohydrate translocation exist: translocation of sucrose through the apoplast and through the symplast. Parallel to the transportation of sucrose, translocation of herbicides occurs in plants [112].

A study on the translocation abilities of six penethylamine derivatives with different K_{ow} and pK_a indicated that strong bases with a pK_a 9.5 and medium lipophilicity (with $\log K_{ow} \sim 2-3$) were better assimilated by roots and translocated to shoots. The assimilation degree significantly decreased with a reduction in pH to 5.0. This can be explained by the fact that amine bases in an acidic medium capture protons (are protonated), acquire charge, and hence their permeability through membranes becomes limited [247].

A study of the assimilation of some fungicides, herbicides and insecticides of different chemical classes by soybean roots and their penetration into the xylem showed that the lipophilicity of the xenobiotic greatly affects these processes and that the maximum concentration of each pesticide in xylem sap is reached at a $\log K_{ow} \sim 3$ [462].

However, despite its lipophilic character, the fungicide morpholine is systemic, i.e., it penetrates and translocates throughout the plant. To explain this phenomenon assimilation and transport of the labeled morpholine fungicides ^{14}C-dodemorph and ^{14}C-tridemorph at different pH have been studied. At pH 5.0 assimilation and translocation rates were insignificant, at pH 8.0 the rates increased by approximately two orders of magnitude. At pH 8.0 the more lipophilic tridemorph accumulated in roots and was moderately rapidly translocated to shoots. Dodemorph accumulated in roots in lesser quantity, but was translocated effectively through the epidermis into the xylem. This picture was similar in 24 h and 48 h incubations. This indicates that assimilation and translocation of toxic compounds in plants maintain a balance in plant cells, at least within limited periods of time [72].

The material presented above allows one to draw the conclusion that higher plants seem to be universal organisms with the potential to absorb contaminants with different chemical structures from soil, water and air. Initially, once absorbed by roots and leaves, the toxic compounds are distributed into the cells of all plant organs by the transpiration stream and assimilate flows, and then are completely degraded (to CO_2) or accumulated in plant cell vacuoles and intercellular space. Because of this ability, plants are already used in different phytoremediation technologies. However, to fully realize the utilization of the ecological potential of plants, involving enzymes deeply degrading organic xenobiotics to carbon dioxide, and, thus, returning carbon atoms to natural circulation, these deep degradation processes should be further studied. The induction mechanisms of these en-

zymes, their overall intracellular distribution and the regulation of their activities seem to be especially promising avenues of further investigation and wide potential application. Another characteristic important for phytoremediation applications is the resistance of plant cells to certain environmental contaminants of diverse structures.

2.2 The action of environmental contaminants on the plant cell

The plant's abilities to absorb, deposit (conjugate), and deeply degrade pollutants, and to mineralize organic and to accumulate inorganic pollutants within its cells determines the ecological potential of the plant. These abilities are the main technological parameters determining the application of plants in novel phytoremediation technologies. To investigate the ecological potential of plants, besides ascertaining their physiological and biochemical characteristics, studies at the level of cell ultrastructure are also very important. Transmission and scanning electron microscopy, in combination with autoradiographic methods, with well-developed techniques of fixation of plant tissues and obtaining ultrathin sections, allows the deleterious effects of toxic contaminants to be revealed at the ultrastructural level, and the fate of toxicants in the plant cell to be followed.

Investigations carried out in different countries have shown that the complex morphological changes and alterations in the main metabolic processes of plant cell elicited by organic pollutants (pesticides, hydrocarbons, phenols, aromatic amines, etc.) are connected with the destruction of the cell's ultrastructural architecture [5, 52, 56, 127, 278, 284, 544, 545]. Often, even partial destruction of the ultrastructure of plant cell organelles promotes the occurrence of various pathological processes, including changes in the intensity of vitally important intracellular processes that initiate cell death.

2.2.1 Changes in cell ultrastructure

The assessment of deviations in cell ultrastructural organization under the action of xenobiotics (environmental pollutants) allows the toxic dose to be determined and serves as one of the indicators for the evaluation of the detoxification potential of each plant species [278, 544]. The sequence and characteristics of the destruction of the plant cell organelles depend on the chemical nature, concentration and duration of the action of the contaminant, the degree of resistance of the plant cell, and other factors [56].

Penetration and movement of contaminants in the cell

Plants absorb organic contaminants from the air and the soil after being dissolved in water and taken up together with nutritional compounds [5, 278]. As discussed in section 2.1, the absorption of pollutants differs between roots and leaves. Plants control the absorption of dissolved xenobiotics. The rhizodermis of young roots absorbs foreign compounds by osmosis. This absorption depends greatly on external factors such as temperature and pH of the ambient nutritional solution and the soil. Other factors that significantly influence the penetration of toxicants into the plant is their molecular mass and their lipophilicity (hydrophobicity), the latter principally affecting the movement of organic pollutants across the plant cell membrane. After passage across the membrane, xenobiotics are distributed throughout the entire plant.

This course of events was experimentally demonstrated in a number of investigations. In various higher plants exposed to different ^{14}C-labeled toxic compounds the label was found in homogenates of roots and leaves [84, 127, 261, 342, 515].

The penetration, movement and localization of contaminants in plant cells have been studied at the ultrastructural level mainly by means of electron microscopy. Electron micrographs of maize root apex cells exposed to ^{14}C-nitrobenzene shows its penetration across the plasmalemma and localization in different subcellular organelles (Fig. 2.11).

Studies of the penetration of ^{14}C–labeled xenobiotics into the cell indicate that labeled compounds at the early stages of exposure (5–10 min) are detected in the cell membrane, in the nuclei and nucleolus (in small amounts), and, seldom, in the cytoplasm and mitochondria. As a result of prolonged exposure, the amount of a label significantly increases in the nucleus, at the membranes of organelles, and in tonoplasts, and further in vacuoles [544]. Thus, xenobiotics with many different structures become distributed in most or all subcellular organelles, but ultimately there is a tendency to accumulate in vacuoles.

The rate and the degree of organic contaminant penetration into the plant cell vary from plant to plant, and depend on the structure of the toxicant and its lipophilicity. Regardless of the plant species, specific lipid bilayer membranes have to be crossed. Studies on the interaction of model pollutants with model lipid bilayer membranes indicate that the rate and degree of penetration depend strongly on the chemical structure of the lipids [395, 396]. Thus, much more attention should be paid to identity of and, possibly, to modifying these crucial lipid components of plants.

Exposure of maize seedlings to a 1 mM solution of [1-^{14}C] phenoxyacetic acid for 10 min resulted in the detection of the herbicide-derived label in the nucleus, nucleolus and vacuoles of the root apex cells. At that epoch, the radioactive label occupied about 3% of the total cell's cross-sectional area. In sunflower, within the same interval, the [1-^{14}C] phenoxyacetic acid label covered over 16% of the apical cells, although unlike maize the contaminant was concentrated in the intercellular space and less was found in the nucleus (Fig. 2.12). Analogously to sunflower, phenoxyacetic acid abundantly penetrates the cellular membrane of pea root apex cells, and is localized in the cytoplasm, nucleus and nucleolus [278].

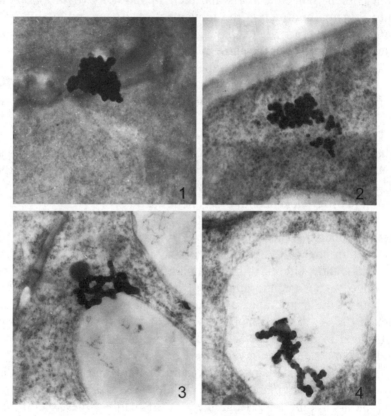

Fig. 2.11. Electron micrographs showing the penetration and movement of ^{14}C-labeled nitrobenzene (0.15 mM) in a maize root apex cell. The xenobiotic penetrated through the plasmalemma (1), moved to the cytoplasm (2), and was thereafter translocated directly into vacuoles (3,4).
1 – x 48 000; 2 – x 36 000; 3 – x 50 000; 4 – x 30 000

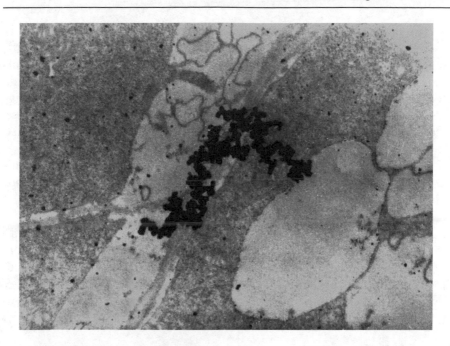

Fig. 2.12. Fragment of a sunflower root cell after 10 min exposure to a 1 mM solution of $[1-^{14}C]$ phenoxyacetic acid. Penetration of the label into different cell organelles is visible. x 50 000

Maize cells were less permeable to labeled $[1-^{14}C]$ 2,4-D than to phenoxyacetic acid. 2,4-D penetrated more easily and deeply into pea and sunflower cells and then was localized in different subcellular organelles. Incubation of different plants with this contaminant for 30 min showed that in maize cells the herbicide was located in the nucleus or vacuoles, and in small amounts (occupying less than 5% of the total cell cross-sectional area) in the cytoplasm and membranes. In sunflower and pea the radioactive label of the same compound occupied more than 30% of intracellular space, and was located in mitochondria, plastids, nucleus and nucleolus [54, 55]. The lowest permeability for contaminants was found in maize (7%), an intermediate value in pea (22%), and the highest in sunflower (50%).

Exposure of maize seedlings to the aromatic amine $[1-6-^{14}C]$ benzidine (at a concentration of 2.2 μM) led after 10 min to penetration of the contaminant into the cells. Within this period the benzidine crossed the cell wall and plasmalemma and was translocated into the nucleus. After 30 min, the contaminant penetrated into the nucleolus, and after 60 min most benzidine was accumulated in the vacuole as a result of the plant defense reaction [545].

The penetration of [1-^{14}C] benzoic acid proceeded far more slowly in maize cells. This compound, at an exposure concentration of 1 mM, reached the nucleus in 1 h. After 24 h, the radioactive label occupied almost the whole cell, i.e. cytoplasm, plastids, mitochondria, endoplasmic reticulum and tonoplasts of vacuoles, but in the nuclei the label was significantly sparser. After 72 h of incubation, the radioactive label was largely accumulated in the vacuoles, while insignificant amounts remained in the other organelles and cytoplasm [84].

According to the above-mentioned results, and supported by data reported elsewhere in the literature [52–56, 277, 544–546], it may be concluded that the penetration mechanisms of contaminants into the plant cell does not follow one sole pattern. The penetration of organic xenobiotics into plant cells depends on plant species, the compounds' chemical nature (lipophilicity), molecular size, temperature, and doubtless other factors whose influence on the process remain to be clarified. On the other hand, the distribution of organic toxicants within plant cells is a far less complex process. The main site of their ultimate localization is the vacuoles, and to a lesser extent mitochondria, plastids, nucleus and nucleolus.

Ultrastructural changes

The general picture of the evolving action of organic contaminants on plant cells during exposure is the following:
- Initially, changes in the configuration of the nucleus become noticeable. Simultaneously inhibition of DNA synthesis [545] takes place. The barrier function of the plasmalemma and its ability to accumulate calcium are damaged. Ca^{2+} concentration in the cytoplasm is enhanced [278, 545] and Ca^{2+}–ATPase activity is inhibited. In cells subjected to the action of environmental pollutants, mitochondria with swollen cristae and a packed matrix (Fig. 2.13) are found, and the plastids are electron-dense and enlarged.
- Prolonged action of contaminants leads to a widening of the cisternae of the endoplasmic reticulum and Golgi apparatus, and vacuolization of the cytoplasm. The size of the cytoplasm is thereby decreased and the periplasmic space concomitantly enlarged. In some cortical cells of the root apices, the number of ribosomes in the hyaloplasm is increased, and the formation of polysomes is observed. Lysis of mitochondria and depletion of ribosomes from the endoplasmic reticulum membranes take place. Multiple contacts between the endoplasmic reticulum and the plasmalemma, plasmadesma, vacuoles, nucleus, and membranes of the mitochondria are detectable. Enhancement of the size of the nucleus and chromatin coagulation has been observed, indicating a disturbance of

the process of DNA synthesis. Nuclei acquire deviant shapes because of the development of many protuberances of the nuclear membrane (Fig. 2.14). In leaf cells, the shape and consistence of the chloroplasts become damaged, the external membrane is not visible, the orientation of the system is disturbed, and the matrix is brightened with large osmiophilic inclusions. In the cytoplasm of the differentiated cells of the root caps that secrete mucus, accumulation of hypertrophied secretory vesicles is visible, most of which remain at the place of their formation or stay connected with the cytoplasm organelles (e.g., mitochondria) instead of being translocated to the periphery and fused with the plasmalemma. Some of these hypertrophied vesicles are fused, forming a large deposit of mucus. Inhibition of the translocation of maturing secretory vesicles towards the cell periphery is often correlated not only with the swelling of vesicles, but also with the disappearance of the normal dictyosomes.
– Longer exposure to environmental contaminants causes extensive destruction of the cell and plant death.

These data have generally been obtained during the study of different contaminants at fixed concentrations. The degree of deviation from normal ultrastructure is highly dependent on the concentration of the xenobiotic. Thus, plants can act as effective remediators only at comparatively low contamination levels, at which no significant changes in cell ultrastructure occur. However, plants subjected to metabolizable concentrations for relatively short periods are in most cases able to recover from small deviations in cell ultrastructure [84] and thus maintain their vital activities.

Low-molecular-mass alkanes and alkenes. To study the effect of low-molecular-mass saturated and unsaturated hydrocarbons on the ultrastructure of leaf cells, maize seedlings were incubated in gas-tight growth chambers filled with an atmosphere containing 25% (by volume) of hydrocarbon for 48 hours [52, 53].

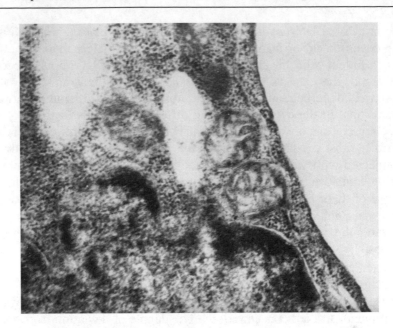

Fig. 2.13. Fragment of a soybean root cell after 10 min exposure to a 1.5 mM solution of nitrobenzene. Mitochondria with swollen cristae are visible. x 25 000

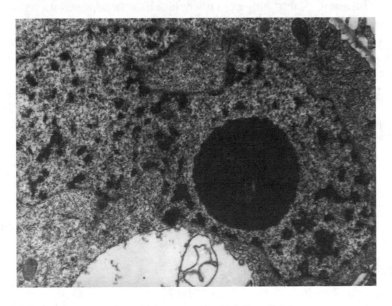

Fig. 2.14. Fragment of a maize root cell after 24 h exposure to a 3,4-benzopyrene solution (1.0 mM). Note the intensive invagination of the nuclear membrane. x 60 000

In the case of methane, the following morphological changes have been observed in the epidermal cells of maize seedling leaves (Fig. 2.15): the chloroplasts were distributed on the periphery of the cell and contained a great amount of starch. Vacuolar bubbles were visible on the external membranes of the chloroplasts. Elongated lamellar membranes were stretched along the whole chloroplast over the longitudinal axis. Large quantities of lipid insertions as well as mitochondria with an electron-dense matrix and widened cristae were concentrated around the chloroplasts. Completely destroyed chloroplasts without double membranes were noticeable in some cells. Separate grains of these chloroplasts were split and located in the brightened matrix of the cytoplasm. The chloroplasts of the cells in the middle part of the maize leaf were crescent shaped. In most cases the chloroplasts were grouped along the cell periphery. In some cells destroyed grains and chloroplasts with lysed internal membranes were visible. The mitochondria had brightened or electron-dense matrices. Cells of the inferior part of the leaf were less affected by the methane action. In these cells, the chloroplasts had a more circular form and accumulation of starch grains was not noticeable. Some of the mitochondria had a brightened matrix [52].

The action of ethane on plant cells differed slightly from the action of methane. In the upper part of the maize leaf, normal chloroplasts with crescent-shaped grains were observed. A small proportion of the chloroplast population was stretched and elongated. Some chloroplasts contained large quantities of starch grains (Fig. 2.16). The chloroplasts in the cells of the middle part of the leaf acquired an elongated shape. They contained in most cases starch grains and only the chloroplast matrix remained noticeable. Swollen membranes surrounded the chloroplasts. Chloroplasts in various stages of destruction were found: with destroyed external membranes, and with dissociated grains in the thylakoids. In these cells the mitochondria were invaginated and contained electron-dense matrices with widened cristae. The cell cytoplasm was brightened. In the lower part of the leaf, ethane did not cause any specific deviations in the shape and structure of the chloroplast, similar to methane. The chloroplasts had clearly visible grains and thylakoids, and the mitochondria had dense matrices with swollen cristae [52].

Damage to the photosynthetic apparatus was revealed under the influence of other gaseous alkanes, i.e., propane, butane and a mixture of alkanes with a composition similar to natural gas (88.7% methane, 6.8% ethane, 2.8% propane, and 1.7% butane) on maize and ryegrass (*Lolium perenne*) seedlings [52, 53].

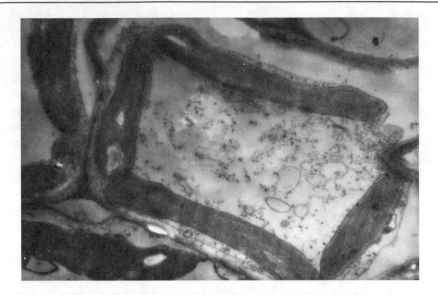

Fig. 2.15. The chloroplasts in epidermal cells of maize leaves (upper part) exposed to methane (25% by volume) for 24 h. The changes in shape of the chloroplasts are visible. x 30 000

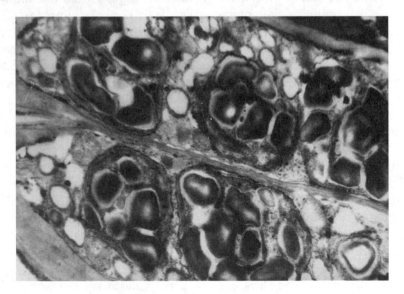

Fig. 2.16. The chloroplasts in epidermal cells of maize leaves (upper part) exposed to ethane (25% by volume) for 24 h. Large quantities of starch grains are visible. x 30 000

Generalizing these results, it is evident that the chloroplasts of the upper and middle parts of the leaf are more sensitive to the action of gaseous alkanes (Fig. 2.15–18). Densitometry analysis of chloroplast thylakoids has shown that after 24 h of incubation of leaves in an alkane atmosphere the average cross-sectional area of chloroplasts increases upon transfer from exposure to methane to butane, and the thickness and distance between the thylakoid membranes decrease.

It seems that one of the main pathological effects characteristic of low-molecular-mass alkanes is the ability to swell chloroplasts and cause morphological and ultrastructural changes at the level of the membrane. The swelling of the thylakoid membranes of chloroplasts, inducing their morphological change, leads to a significant decrease in the matrix of the chloroplasts that finally adversely affects the normal range of biochemical processes associated with the plant photosynthetic apparatus.

Incubation of pea seedlings in an atmosphere containing 25% (by volume) propylene or butylene for 72 h leads to partial plasmolysis of the plastids and mitochondria in the upper part of the leaf [52]. Unlike with the alkanes, in this case reduction of the chloroplast size, destruction of the lamellar systems and accumulation of the starch grains take place. Mitochondria are swollen, their matrices become dense, and cristae are widened. In cells of the middle part of the leaf, the destructive action of propylene is enhanced. Higher accumulation of starch is seen in plastids, in some cells chloroplasts are destroyed and dispersed in the cytoplasm, where some separate grains and thylakoids are observed. Mitochondria are affected at the initial stage of destruction—the loss of matrix density makes them appear brighter and the majority of cristae disappear. The inferior part of the leaf undergoes significant changes, mainly at the level of chloroplasts and mitochondria. The cytoplasm becomes brighter, the nucleus frayed, and the membrane complex less distinct.

Low-molecular-mass alkenes induce somewhat similar changes in cell structure. However, alkenes affect the inferior parts of the leaf more, unlike alkanes that toxify the upper and middle leaf parts. In both cases chloroplasts and mitochondria are damaged most.

The photosynthetic apparatus appears to be the most sensitive to the action of other gaseous contaminants as well [182, 454].

Aromatic ring-containing organic contaminants. The action of structurally different contaminants on the plant cell ultrastructure, notably compounds containing an aromatic ring, has also been examined.

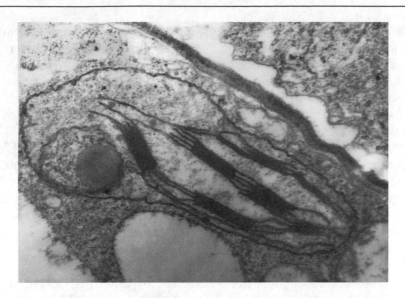

Fig. 2.17. The chloroplasts in cells of maize leaves (lower part) exposed to ethane (25% by volume) for 24 h. Large quantities of starch granules are visible. x 30 000

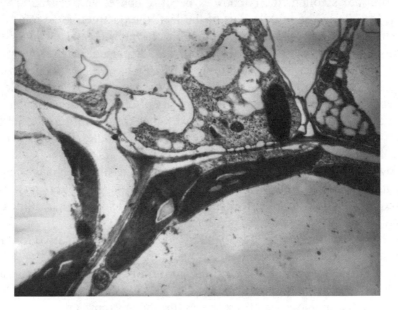

Fig. 2.18. The chloroplasts in cells of maize leaves exposed to a mixture of methane (88.7%), ethane (6.8%), propane (2.8%) and butane (1.7%) for 24 h. Elongation of the chloroplasts, chaotic dislocation of grains and cell destruction are visible. x 30 000

As a result of the action of [1–6-^{14}C] benzene on the ultrastructural organization of the cells of perennial woody plant leaves, first of all pathological changes are observed in the photosynthetic apparatus. This is prominently expressed in the disorganization of the chloroplast-lamellae-grains complex and the appearance of osmiophilic inclusions in the chloroplasts (Fig. 2.19) [278]. In plants known to be highly resistant to benzene action, such as lime, maple, silver fir (*Abies* sp.), poplar, Norway spruce (*Picea abies*), common fir (*Picea* sp.), walnut (*Juglans regia*), plane (*Platanus orientalis*), cypress (*Cupressus sempervirensy*), and ash (*Fraxinus excelsior*), the photosynthetic apparatus is resistant to the action of benzene.

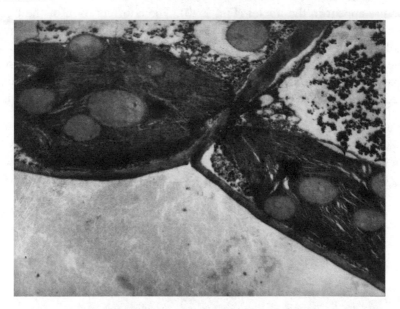

Fig. 2.19. A maple leaf fragment exposed to [1–6-^{14}C] benzene (0.1 mM) for 24 h. Note the changes in structure of the photosynthetic apparatus and accumulation of inclusions. x 60 000

A low concentration of benzene vapor (0.04 mM) causes only insignificant changes in the cell ultrastructure of leaves of 7-day-old bean seedlings, resulting in brightened fragments in the chloroplast matrix. Increasing the benzene concentration five-fold (to 0.2 mM) causes a destructive effect only in chloroplasts, i.e., the disorientation of lamellae and thylakoids. Complete cell destruction occurs at exposure to 0.4 mM benzene: in chloroplasts the internal membrane system is damaged and the matrix becomes electron-dense, the cell wall is thickened, myelin and osmiophilic inclusions are observed in the periplasm and in the vacuoles, and the mitochondria become electron-dense.

The aromatic nitroderivatives nitrobenzene, *o*-nitrophenol and 2,4-dinitrophenol (at concentrations of 1 mM) induce total destruction of the cell ultrastructure in the upper and lower parts of leaves. Benzene, phenol, *o*-nitrophenol, and *o*-cresol induce pathological destruction only in the lower parts of the leaves. It appears that these different toxicities of xenobiotics are determined not only by the presence of the reactive functional groups, but also by their location within the xenobiotic molecule [337].

After penetration of the PAHs (1,2-benzanthracene and 3,4-benzopyrene) into the roots, the first signs of destruction of the cell ultrastructure appear in the nuclei (Fig. 2.20). The configuration of the nuclear membrane is significantly changed – the nucleus becomes invaginated. Small concentrations (0.1 mM) of 3,4-benzopyrene do not have a clear pathological effect on the cell structure [56]. Supposedly, due to deep oxidative transformations, the 3,4-benzopyrene degradation products are inserted into metabolic processes and further oxidation to carbon dioxide occurs [278].

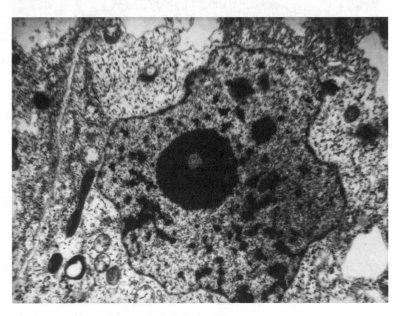

Fig. 2.20. A fragment of a miaze root cell exposed to a 1 mM solution of 3,4-benzopyrene for 24 h. x 60 000

Upon a further increase in the concentration of 1,2-benzanthracene and 3,4-benzopyrene (up to 1 mM) coagulation of chromatids is observed, which indicates impairment of the DNA synthesis. The mitochondria lose their internal content and become brightened. Afterwards, total cell destruction takes place at a concentration of 10 mM. Plastids appear more resistant to PAHs than other organelles.

The effect of different benzidine concentrations on the ultrastructural organization of maize root apex cells has been investigated [545]. It was demonstrated that benzidine at a concentration of 0.22–2.2 µM did not cause noticeable ultrastructural changes. When the benzidine concentration was increased up to 2.2 mM several pathological changes occurred, finally destroying the cell. Autoradiographic studies have demonstrated the inhibition of DNA synthesis, and cytochemical methods have revealed the inhibition of Ca^{2+}-ATPase.

Another example of the strong dependence of a pathological action on increasing contaminant concentrations is the action of the herbicides dinitro-o-cresole (DNOC) and 2,4-D on the photosynthetic apparatus of vine leaf cells. These herbicides cause the loss of the normal organization of the surface of vine leaves. In cells of such leaves smoothing of the mitochondrial cristae and a reduction in stomatal size is observed. Such structural changes in the epidermis finally lead to the loss of leaf elasticity [52].

In the root cortical cells of soybean exposed to a 0.5 mM solution of labeled [1-^{14}C] TNT for 5 days, the contaminant that has penetrated into the cell is observable as electron-dense matter in the cell walls, endoplasmic reticulum, mitochondria, plastids, nucleolus and vacuoles (Fig. 2.21). This concentration of TNT leads to the total cell destruction in maize roots. In soybean leaves the labeled TNT is detected in the cell wall, chloroplasts and vacuoles. In maize leaf cells, the distribution of the label is similar. Comparative analysis creates the basis for concluding that soybean is more resistant to TNT action than maize.

Attention should be paid to the localization of TNT on membrane structures participating in the transport of reducing equivalents (membranes of the endoplasmic reticulum, mitochondria, and plastids). Supposedly, TNT transformation proceeds in these subcellular organelles.

Effect of inorganic gases on plant cell ultrastructure. The study of plant responses to SO_2 exposure has revealed acute and chronic injury that results in chlorosis and subsequent necrosis due to the destruction of chlorophylls and finally lysis of the chloroplasts. It has been documented that ultrastructural characteristics of leaves are affected prior to any visible injury. Electron microscopic examination of SO_2-fumigated plant-attached leaves of pigeon pea (*Vicia faba*) revealed that the chloroplast thylakoids started to swell while the photosynthesis rate was drastically reduced [390].

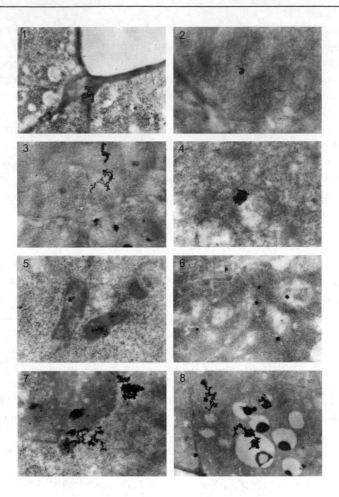

Fig. 2.21. Root cortical cells of soybean seedlings, exposed to 0.5 mM [1-^{14}C] TNT for 5 days.

1 – Label in cell wall. x 20 000
2 – Label in contact with the outer membranes of the mitochondria. x 36 000
3 – Label in the mitochondria, on the plasmalemma, in the endoplasmic reticulum. x 28 000
4 – Label in the mitochondria. x 48 000
5 – Label in the plastids. x 36 000
6 – Label in the mitochondria and endoplasmic reticulum. x 20 000
7 – Label in the nucleus and nucleoli. x 28 000
8 – Label in the vacuoles, plastids and mitochondria. x 18 000

Varying responses of leaf cuticles to pollution stress have also been demonstrated [29, 431]. Manninen and Huttunen [325] using electron microscopy observed that the epicuticular wax structure of Scots pine needles was badly degenerated in trees in close vicinity to an oil refinery in southern Finland. The degree of destruction of the surface wax of the needles increased with increasing sulfur content of the needles.

A review of plant cells and tissues as bioindicators of environmental pollution indicated that ultrastructural responses to O_3 are not specific. The effects of peroxyacetyl nitrate and O_3 are often indistinguishable [506].

Degeneration of the needle wax layer in response to SO_2 has been demonstrated [325], but O_3 does not significantly alter the needle wax layer in Norway spruce [23].

Pollutants often occur in mixtures. It is, therefore, necessary to investigate the effects of pollutant mixtures on plants. Responses that allow the identification of pollutant mixtures would be extremely useful in bioindication. Obviously, investigations of possible synergies between different pollutants require extensive studies if the different responses to different pollutant combinations are to be adequately characterized. From the limited data presently available it is apparent that plants can be far more severely affected by mixtures of pollutants than by the same pollutants applied individually [226]. For instance, closure of stomata occurs at lower concentrations of O_3 if O_3 and SO_2 are present together [94].

Studies involving contaminant mixtures have been carried out mostly for mixtures of O_3 and SO_2 [167, 334]. To a lesser extent SO_2 and NO_2 combinations have been investigated [263] and sometimes all three pollutants in combination have been studied [57]. Studies have also been conducted with metal and gaseous pollutant mixtures [126, 142, 265].

Ultrastructural demonstration of structure-function changes directed to xenobiotic detoxification

Plants significantly differ in their ability to assimilate organic pollutants that penetrated into the cytoplasm. For the first 30 min xenobiotics penetrate into and accumulate in the subcellular organelles. Simultaneously, the induction of specific enzymes that participate in further oxidative transformations of the xenobiotics takes place [266, 288]. All toxic compounds investigated so far changed the plant cell structure to a greater or lesser extent. Despite the fact that at the lower, so-called metabolic, concentrations the normal cytological status did not change, it should be taken into consideration that even in this case some deviations in cell ultrastructure (e.g. widening of periplasmic space, diminution of plasmodesmata, increase in the volume of the endoplasmic reticulum etc.) take place.

Cytochemical investigations have shown that exposure of maize seedlings to 0.22 mM benzidine for 24 h caused inhibition of Ca^{2+}-ATPase and an increase in the concentration of free calcium in the cytoplasm resulting from the release of calcium from the membrane system. These data suggest that ultrastructural changes are caused by the disturbance of cell calcium homeostasis, resulting in the deregulation of the Ca^{2+}-dependent metabolic processes [545].

Attention should be paid to the processes that promote the detoxification of pollutants after entrance and their removal from the cell. Among such processes, deposition of xenobiotics in the vacuole must be emphasized. This phenomenon allows the cell to resist the destructive action of the toxicant at least temporarily. Sending the xenobiotic to the vacuole excludes it from interfering with normal cell metabolism.

Partially transformed toxicants are also stored in the vacuole. These are conjugates of xenobiotics or intermediate products of their transformation with intracellular compounds (proteins, peptides, low molecular weight sugars etc.). For instance, after the penetration of labeled 2,4-D into the root cells of barley *(Hordeum vulgare)* seedlings, conjugates were detected in the vacuoles, and among these conjugates 80% were O-β-D-glycosides of the herbicide metabolites [79].

Usually, the quantity and size of the vacuoles are significantly increased under the action of the toxicant. In addition to intensification of vacuolization, the fusion of some small vacuoles and formation of larger organelles, often occupying most of the cell cross-sectional area, is visible. However, as soon as the cell is given a chance, the process of removing the toxic residues from the vacuoles to the extracellular and subsequently to the intercellular spaces begins. This phenomenon is observed after terminating the exposure of the plant to the toxicant. In such cases the periplasmic space of the cell is appreciably widened. The smoothing of the rough endoplasmic reticulum begins, then the cisternae of the smooth endoplasmic reticulum connect with vacuoles through which part of the conjugates are excreted from the cell. Besides, fragmentation of the cisternae of the endoplasmic reticulum takes place, which proceeds in the form of vesicles that are translocated to the periphery of the cell. In the final phase of the process of exocytosis, the secretion process is activated. The observed formation of multiple contacts of vesicles with the plasmalemma confirms the hypothesis. Fusion of the membranes of vesicles and plasmalemma proceeds via the participation of Ca^{2+}-binding sites. As a result, the vesicle contents (conjugates of pollutants) are removed to the periplasmic space (Fig. 2.22 b, c, d).

Fragments of the smooth endoplasmic reticulum participate in another simple process of secretion of the vesicle content beyond the cell. In this case, the channels connecting the vacuoles with the plasmalemma for the translocation of the toxic remains from the vacuole to the intercellular space are formed by means of smooth membrane fragments (Fig. 1.37 a).

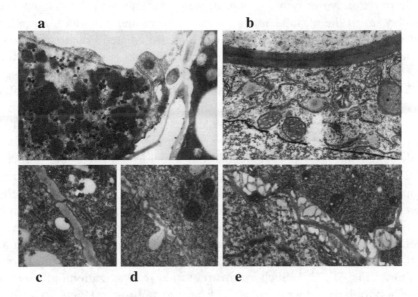

Fig. 2.22. Fragments of root apex cells of maize seedlings, which after exposure to 2,4-D (concentration 1.4 µM; exposure time 96 h) were transferred to a medium lacking 2,4-D.
a) Contacts between the endoplasmic reticulum and the vacuoles. x 20 000
b) Fragmentation of a granulated endoplasmic reticulum. x 30 000
c) Contacts between vesicles and the plasmalemma. x 20 000
d) Fusion of vesicular membranes with plasmalemma. x 25 000
e) Membrane fragments accumulated in the periplasm. x 25 000

Often a large number of ribosomes is visible in plant cells affected by pollutants. This phenomenon points to an increase in protein biosynthesis. Electron microscopy of ultrathin sections of soybean and maize roots apices, exposed to nitrobenzene at different concentrations, clearly showed the appearance of cells darkened by numerous ribosomes [546]. At lower concentrations of nitrobenzene (1.5×10^{-5} M) such cells were found only in the apex of maize roots. In soybean cells, rich in reserve protein complexes, the number of ribosomes increased only at a pollutant concentration of 1.5×10^{-3} M, when the store of proteins was consumed. The increase in protein biosynthesis could be explained by induction of the enzymes participating in the contaminants' intracellular detoxification and supply of the diminished amount of protein during the conjugation process. Histochemical and biochemical analyses showed that concurrent induction of enzymes important for detoxification (peroxidases, cytochrome P450-containing monooxygenases and phenoloxidases) takes place. The content of these oxidative enzymes is significantly enhanced in the cell wall, on membranes of the plasmalemma, in the endoplasmic reticulum, tonoplasts and vacuoles, i.e. where suitable conditions are created to detoxify the organic contaminants in preparation for their removal from the information-processing and energetic centers of the cell [288].

Presumably, cells attempt to minimize the destructive action of pollutants via their deep degradation, and this is expressed in the strong induction of enzymes participating in the detoxification processes, and in the creation of optimum conditions for consecutive and effective functioning of these enzymes. Organic toxicants that undergo oxidative and reductive transformations often induce multiple contacts between organelles (mitochondria, endoplasmic reticulum, and plastids) equipped with a redox (electron transfer) chain for electron transport on membranes. For instance, in maize root cells exposed to nitrobenzene, the contacts between the endoplasmic reticulum and the mitochondria were quantified and many mitochondria appeared to be surrounded by membranes of the endoplasmic reticulum (Fig. 2.23). Such ultrastructural reorganization allows the mitochondrial and microsomal electron-transporting systems to provide reducing equivalents to the cytochrome P450-containing monooxygenase system located on the endoplasmic reticular membranes. Cytochrome P450 uses these electrons for the activation of molecular oxygen and the hydroxylation of xenobiotics, which is the rate-limiting step of the whole detoxification process. It is, therefore, clear that the consumption of electrons provides the main reason for the close juxtapositioning of mitochondrial and microsomal membranes in plant cells affected by xenobiotics. This phenomenon is observed in both animal [432] and in plant [197] cells, and is known as the "mitochondrial control" of xenobiotic oxidation.

Similar ultrastructural changes leading to the contacts of membrane structures are observed in plant cells exposed to TNT. In this case, other providers of reducing equivalents, i.e. plastids, are in contact with the endoplasmic reticulum, together with mitochondria. In this case, electrons are needed not for oxidation, but for reduction of the nitro-groups of TNT, leading to the formation of the less toxic amino-derivatives of toluene.

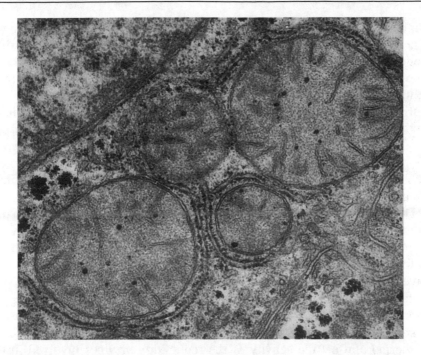

Fig. 2.23. Fragment of root apex cells of maize seedlings exposed to a 0.15 mM solution of nitrobenzene (for 24 h). A large number of mitochondria surrounded by membranes of the endoplasmic reticulum is observed. x 60 000

Thus, it is evident that after environmental contaminants penetrate into the plant cell, changes in structure and function take place as a result of contaminant toxicity and cell reorganization directed to pollutant detoxification.

2.2.2 Changes in the activities of regular metabolism enzymes

Organic pollutants which have penetrated into plant cells cause significant changes across the whole range of intracellular metabolic processes. This phenomenon is in agreement with biochemical logic and should be no surprise: degradation of toxicants that entered the cell requires the full mobilization of the cells' internal metabolic potential. The cell is fighting foreign and at the same time toxic compounds with all accessible means. This is firstly manifested in the activation of inductive processes directed towards the synthesis of enzymes and enzymatic systems participating in xenobiotic detoxification. As a result of the progressive oxidation of the xenobiotics, standard cellular intermediates are formed, which then get entrained in the general metabolic cycle and are oxidized ultimately to carbon dioxide. Degradation of organic pollutants leading to the formation of standard metabolites is in full accordance with the biological principle of economy of cellular processes – one of the most important underlying principles of life. During the detoxification (especially oxidation) processes a significant amount of endogenous cell energy, as well as metabolites vitally important for the cell, are expended. Presumably, the great majority of cell enzymes are mobilized and directly or indirectly involved in the xenobiotic degradation process. Conceptually it is the vital equivalent of le Châtelier's principle in operation.

Despite the fact that biochemical processes accompanying the detoxification process in plants are not well investigated, in the literature many examples exist indicating that the activities of the enzymes participating in different regular cellular processes are also influenced by xenobiotics that have entered the cell. For instance, in alfalfa inhibition of glutamine synthetase activity (up to 50%), a simultaneous stimulation of glutamate dehydrogenase activity (40%) took place as a result of phosphinotricine exposure [24]. Toxicants that have penetrated into the plant cell may furthermore affect the activity of the regulatory enzymes involved in the tricarboxylic acid (TCA or Krebs) cycle and in the process of oxidative phosphorylation. As a consequence, the processes of biosynthesis of ATP and other energetically important adenosine nucleotides (ADP, AMP and GDP) are affected [24].

The study of plant responses towards environmental pollutants is important for the correct understanding of the possibilities of using plants as bioindicators of pollution. Plant responses to environmental pollutants have initially been studied in connection with gaseous pollution. The majority of published data with regard to gaseous air pollution concerns SO_2. However, in response to decreasing atmospheric levels of SO_2, since the 1970s, the focus has increasingly turned to NO_x, NH_3 and O_3.

Biochemical, physiological, and structural bioindication methods for higher plants have been developed. However, plant responses are often not specific and careful assessment is required to identify cause and effect relationships.

Plant responses depend on development stage, age and nutritional status of the plants [280]. Some air pollution effects on plants have been reported (dose responses and gas absorption) [230, 536]. For example, inhibition of photosynthesis and disruption of chloroplast metabolism as a result of SO_2 exposure were observed under controlled conditions [518]. The chlorophyll content was lower in one-year-old needles of damaged spruce trees in comparison with those of healthy trees at three different sites in northern Germany [193].

Changes in stomatal response, chlorophyll level, photosynthesis and metabolite content were demonstrated in different plant species exposed to ozone. Despite the variety in responses towards ozone at a 200 ppm concentration, most plants close their stomata [94]. Pine seedlings exposed to 0.12 ppm O_3 for 7 h per day during 12 weeks showed a 16% reduction in photosynthesis in comparison with plants exposed to charcoal-filtered air [474]. The chlorophyll concentration decreased in tomato plant leaves exposed to O_3 at levels that do not cause any visible damage [163, 470, 496], and reductions in chlorophyll content prior to visible injury were also reported. According to these data, chlorophyll level could be considered as a measure of leaf damage but not as a specific indicator of pollutant. Plant photosynthesis and stomatal conductance have been demonstrated to be relatively tolerant to NO_X exposure [94].

Several investigations [431] have demonstrated metabolic changes such as changes in amino acid, polysaccharide, and ATP/ADP ratios in plants caused by air pollutants, in particular by SO_2. Changes in the production of phenolics (secondary metabolites) and sugars by willow species, indicating serious deviations in plant metabolism in response to 0.11 ppm SO_2 fumigation for three weeks, were also reported [260].

Fumigation of Aleppo pine with O_3 resulted in a delayed rate of ethane emission, accumulation of polyamines, and an increased pool of reduced glutathione and ascorbate in current year needles [524].

The majority of air contaminants affect enzymes [431]. Ozone fumigation reduced light regulation of ribulose bisphosphate carboxylase activity in barley leaves [322].

Enzyme activity has been used as a biochemical stress bioindicator of air pollutants. According to experiments carries out in Germany, acid phosphatase and peroxidase activities in needles of healthy Norway spruce trees are generally lower than in damaged trees [193]. However, the investigators could not relate these effects to stress-specific factors. Glutathione reductase and ascorbate peroxidase activities in red spruce (*Picea rubers*) needles were enhanced by exposure to acidic mists [77]. The possible role of glutamate dehydrogenase in the adaptation of plants to ammonia assimilation in relation to air pollution has been discussed [434], and it has been suggested that glutamate dehydrogenase activity may be used for the bioindication of ammonia. Plants are often subjected to mixtures of pollutants, which may produce synergistic, additive or antagonistic responses. For example, the threshold limit of tobacco Bel W3 to O_3 exposure is reduced at low concentrations of SO_2 [226].

Special investigations were carried out by the present authors to reveal the effects of different hydrocarbons and aromatic compounds on key enzymes of general plant metabolism, namely glutamine synthetase, glutamate and malate dehydrogenases. For these studies, agricultural plants – monocotyledonous maize and dicotyledonous kidney bean, as well as ryegrass, a perennial monocotyledonous herb – were selected. Ryegrass is characterized by high resistance against several environmental pollutants and by the ability of intensive absorption of organic contaminants and accumulation of heavy metals [52].

In leaves of maize, ryegrass and kidney bean seedlings, exposed to atmospheres containing methane, ethane, propane and butane vapors (10% by volume) in gas-tight chambers for 5 days, significant inhibition of glutamate dehydrogenase (75%) and activation of glutamine synthetase and malate dehydrogenase took place (by 60 and 45%, respectively). As for pentane, in leaves exposed to pentane vapor (5% by volume), inhibition of all enzyme activities has been detected [34].

While aromatic compounds – nitrobenzene and benzoic acid – have a concentration-dependent effect, namely stimulation at lower (1 mM) and inhibition at higher (10 mM) concentrations after transfer of the plants from contaminated to noncontaminated medium, restoration of enzyme activities took place and after 48 h control levels were regained [84]. Changes in the cell ultrastructural organization were also reversible.

The effect of some other aromatic compounds – benzene and 3,4-benzopyrene – on the main metabolic enzymes of maize and kidney bean was studied in more detail, in particular the influence of different concentrations and exposure duration as well as the presence and absence of light. In maize roots, grown under ordinary illumination under the influence of benzene and 3,4-benzopyrene (10 mM, exposure time 72 h) an increase in activities of all enzymes studied took place. The highest increase in activity was observed for glutamate dehydrogenase – 5-fold in the case of benzene and 3-fold in the case of 3,4-benzopyrene. In etiolated seedlings inhibition of glutamine synthetase activity and a 50% increase of glutamate dehydrogenase activity under the influence of 3,4-benzopyrene was noted. Upon an increase in the concentration of the aromatic compounds from 5 to 10 mM, a further (approximately 50%) increase of glutamate dehydrogenase activities both in roots and leaves of maize seedlings was observed. More clear changes in the activities of the studied enzymes were observed in ryegrass seedlings.

The picture was different in case of kidney bean – there, a slight (20%) increase of glutamate dehydrogenase activity in roots at high concentrations (10 mM) of benzene and stimulation of malate and glutamate dehydrogenase activities were observed in leaves.

Not the least important factor is the duration of plant exposure to aromatic compounds. Experiments carried out in this direction have demonstrated that at shorter exposures (2 h) in leaves and roots of maize seedlings none of the studied enzyme activities changed. Upon increasing the exposure duration (24 h) an insignificant stimulation of glutamate dehydrogenase activity both in leaves and roots of maize seedlings was observed. The picture drastically changed at relatively longer (72 h) exposures. In leaves of maize seedlings benzene causes inhibition of glutamate dehydrogenase activity by 40%, malate dehydrogenase by 72%, and glutamine synthetase by 32%. On the contrary, a significant 4-fold stimulation of glutamate dehydrogenase activity and comparatively low stimulation (22%) of malate dehydrogenase were observed in roots.

These studies indicate that aromatic compounds at the given concentrations are toxic for plants upon continued exposure. Evidence is provided by the changes in enzyme activities of regular metabolism as well as in the ultrastructural organization of the cell (Fig. 2.19–20). At comparatively high concentrations of benzene and 3,4-benzopyrene (10 mM) changes are expressed more profoundly. The response of plants to the toxic action of aromatic compounds is expressed via the activation of enzyme systems providing the cell with energy.

Thus, high concentrations of benzene and 3,4-benzopyrene are critical for the studied plants. However, upon prolonged exposure, plants elaborate a defense mechanism against the toxic effect of aromatic compounds that is expressed via ultrastructural reorganization of the cell and mobilization of energetic resources. The increase in glutamate dehydrogenase activity and number of mitochondria, as well as an intensification of their contacts with chloroplasts and endoplasmic reticulum, is evidence for this (Fig. 2.23).

Based on the above-mentioned results, we suggest that investigations of this type point to a relationship between the penetration of organic contaminants into the plant cells and changes in the activities of enzymes participating in energy and nitrogen metabolism. This relationship depends strongly on xenobiotic type and concentration in the cell.

The effect of organic contaminants on plant cells is complicated and can not be easily summarized. However, despite the scarcity of experimental data, it can be suggested that most subcellular organelles and key enzymes of general metabolism are involved in the transformation process of xenobiotics. Resistance against toxic action of xenobiotics consists in the mobilization of all cell resources.

Elucidation of the relationship between detoxification and regular cell metabolism would enable a better understanding of the mechanism of their coordination. The revelation of a plant response, expressed in changes in cell metabolic processes and key enzyme activities, characteristic of each plant species, would serve as an indication for the objective evaluation of plant vitality in polluted environments. Such investigations are important also for understanding plant rehabilitation processes and elaboration of a proper strategy to cultivate agricultural crops on contaminated soils with a guarantee of a safe (for the ultimate consumer) product.

As described in several chapters of this book, despite the fact that the role of plants in the remediation of the environment has been investigated, and participation of key enzymes in the degradation of organic contaminants has been demonstrated, little is known about the processes taking place in the plant cell after penetration of the contaminants, especially concerning the mechanisms determining cell adaptation and survival. The following questions are currently pertinent:

- How is the toxic effect of contaminants on general metabolic processes expressed?
- Are the changes in the plant cell taking place due to environmental contaminants reversible when the contaminant concentration is lowered?
- What are the limits to reversibility of deviations provoked by the action of contaminants?
- How does the rehabilitation process proceed?

Existing knowledge and information do not allow all these questions concerning the action of contaminants on plant cell structure and function to be answered, but clearly indicate that further comprehensive study of the whole process starting with contact of the contaminant with the plant cell and following rehabilitation would help to create the knowledge base for a successful strategy of remediation and prevention of environmental pollution.

We suggest that many detoxification processes in higher plants are closely related to cell metabolism, and not only the absence of particular enzymes or enzymatic systems are rate limiting in the remediation process, but also the cell energetic potential, which determines the limits of additional cell activity.

3 The fate of organic contaminants in the plant cell

3.1 Transformation of environmental contaminants in plants

Nowadays there are many experimental data demonstrating that plants are able to activate a definite set of biochemical and physiological processes to resist the toxic action of environmental contaminants, namely:
- Excretion.
- Conjugation of environmental contaminants with intracellular compounds and further compartmentation of conjugates.
- Degradation of environmental contaminants to common cell metabolites, and finally to carbon dioxide.

Plants totally or partially detoxify environmental organic contaminants entering their cells. These features allow plants to be used in several applications to remove or at least significantly decrease contaminant levels [151].

The main pathways of organic contaminant transformation in plant cells are presented according to Sandermann's "green liver" model [428] in Fig. 3.1.

The simplest pathway for removing organic contaminants from the cell is excretion. The essence of excretion is that the toxicant molecule avoids chemical transformation and is excreted from the plant by translocation through the apoplast. This pathway of contaminant elimination is rare and takes place only at high concentrations of highly mobile (phloem-mobile or ambimobile) compounds. From an ecological point of view the shortcoming of excretion is that the contaminant does not undergo detoxification and returns to the environment with full retention of its toxic features.

In the majority of cases, environmental pollutants absorbed by plants penetrate into the cells where they are subjected to enzymatic transformations leading to the decrease in their toxicity. Nowadays three successive phases of transformation of xenobiotics are considered to be significant [428]:

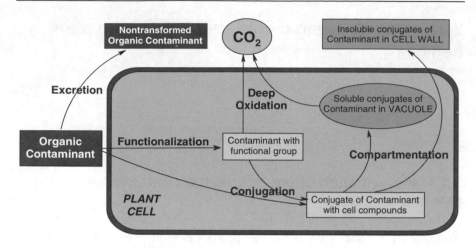

Fig. 3.1. The main pathways of organic contaminant transformation in plant cells

Phase I: Functionalization is a process whereby a molecule of a hydrophobic organic xenobiotic acquires a hydrophilic functional group (hydroxyl, amino, carboxyl, etc.) as a result of enzymatic transformations (oxidation, reduction, hydrolysis, etc.). The polarity and reactivity of the toxicant molecule is enhanced by the acquisition of this functional group. This promotes an increase of its affinity to enzymes, catalyzing further transformation (conjugation or further oxidation). Xenobiotic oxidative degradation proceeds to common cell metabolites and mineralization to CO_2. By this pathway the plant cell not only fully detoxifies the xenobiotic but also utilizes its carbon atoms for intracellular biosynthetic and energetic needs. All these transformations together form the detoxification process. Complete xenobiotic degradation in a plant cell is accomplished only at low, metabolic, concentrations of environmental contaminants, and it requires time. At high concentrations of contaminants full mineralization of xenobiotics is usually not achieved. Typically only a small amount of toxicant present in the cell is mineralized, and the rest undergoes conjugation.

Phase II: Conjugation is a process whereby a xenobiotic is chemically coupled to cell endogenous compounds (proteins, peptides, amino acids, organic acids, mono-, oligo- and polysaccharides, lignin, etc.) by formation of peptide, ether, ester, thioether or other bonds of a covalent nature. Intermediates of xenobiotic transformations or xenobiotics already bearing functional groups capable of reacting with intracellular endogenous compounds are susceptible to conjugation. The formation of conjugates leads to the enhancement of the hydrophilicity of organic contaminants, and consequently to an increase in their mobility. Such characteristics simplify further compartmentation of the transformed toxic compounds. In conjugated form a xenobiotic is kept apart from vital processes in the plant cell and is therefore rendered harmless for the plant [60].

Although conjugation is one of the most widely distributed pathways of plant self-defense against the toxicity of absorbed xenobiotics, it can not be assumed that the process is energetically and physiologically advantageous for the plant [60]. Conjugation leads to the depletion of compounds of significant importance for the cell and their resulting deficiency decreases the capacity of plant resistance to prolonged contamination as well as to changes in environmental conditions. Unlike deep degradation, conjugation does not lead to complete detoxification of the xenobiotic, which preserves its basic molecular structure and hence only partially loses its toxicity. When plants are transferred from a contaminant-containing environment to a contaminant-free medium, gradual degradation of deposited conjugates and mineralization of the toxicant residues take place [84].

Conjugation is not the most successful pathway of xenobiotic detoxification from an ecological point of view. Plant remains, containing the conjugated contaminants, actually become the toxicant carrier. Typically 70% or more of the absorbed xenobiotics are accumulated in plants in the form of conjugates [288]. This must be taken into account when considering the ultimate ecological fate of xenobiotics. Conjugates of toxic compounds are especially hazardous upon entering the food chain: enzymes of the digestive tract of warm-blooded animals can hydrolyze conjugates and release the xenobiotics or products of their partial transformation, which in some cases, due to increased reactivity, are more toxic than the parent xenobiotic. Therefore, it is highly desirable that plants used in phytoremediation have a large capability for the deep enzymatic degradation of xenobiotics. The selection of such plants, and the promotion of gene expression of enzymes participating in plant detoxification processes, are strategies currently followed by modern phytoremediation technologies.

Phase III: Compartmentation is in most cases the final step of conjugate processing. In this phase temporary (short or long term) storage of conjugates in defined compartments of the plant cell takes place. Soluble conjugates of toxic compounds (coupled with peptides, sugars, amino acids etc.) are accumulated in vacuoles, while insoluble conjugates (coupled with protein, lignin, starch, pectin, cellulose, xylan and other polysaccharides) are moved out of the cell via exocytosis and are accumulated in the apoplast or cell wall [428]. The step of compartmentation is analogous to mammalian excretion, essentially removing toxic compounds from metabolic tissues [60]. The major difference between detoxification processes in mammals (and other animals) and plants is that plants do not have a special excretion system for the removal of xenobiotic conjugates from the organism. Hence they use a mechanism of active transport for the removal of the toxic residues away from the vitally important sites of the cell (nuclei, mitochondria, plastids, etc.). This active transport is facilitated and controlled by the ATP-dependent glutathione pump [327]. This process is also termed "storage excretion" [87, 289, 326].

The described pathway of toxic compound processing, i.e., functionalization → conjugation → compartmentation, is illustrated by the degradation of representatives of organochlorine pesticides. For instance, the herbicide 2,4-D is hydroxylated, then conjugated with glucose and malonyl residues, and then is stored in vacuoles (Fig. 3.2) [426].

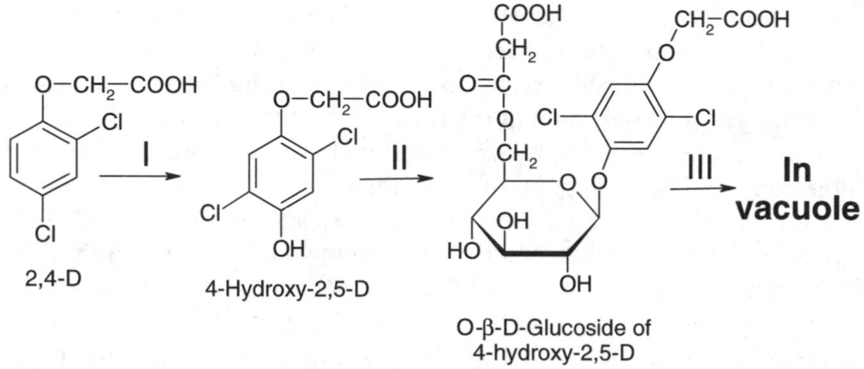

Fig. 3.2. 2,4-D transformation for deposition in vacuoles

The insecticide DDT acquires a carboxyl group via primary oxidation, and then easily forms an ester with glucose, enabling it to be stored in vacuoles according the scheme presented in Fig. 3.3 [428].

The biocide 2,3,4,5,6-pentachlorophenol easily forms the soluble β-D-glucoside and O-malonyl-β-D-glucoside conjugates, which translocate and accumulate in vacuoles. However, if it is hydroxylated, upon which the pentachlorophenol acquires a second hydroxyl group, this intermediate conjugates with lignin, forming an insoluble compound, which is removed from the cell and stored in the cell wall (Fig. 3.4) [428].

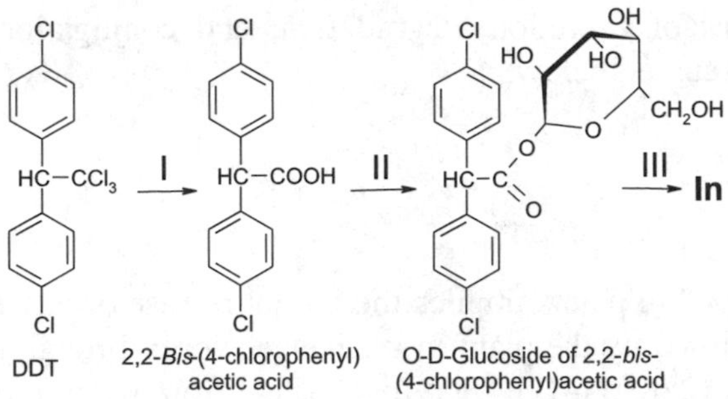

Fig. 3.3. DDT transformation for deposition in the vacuole

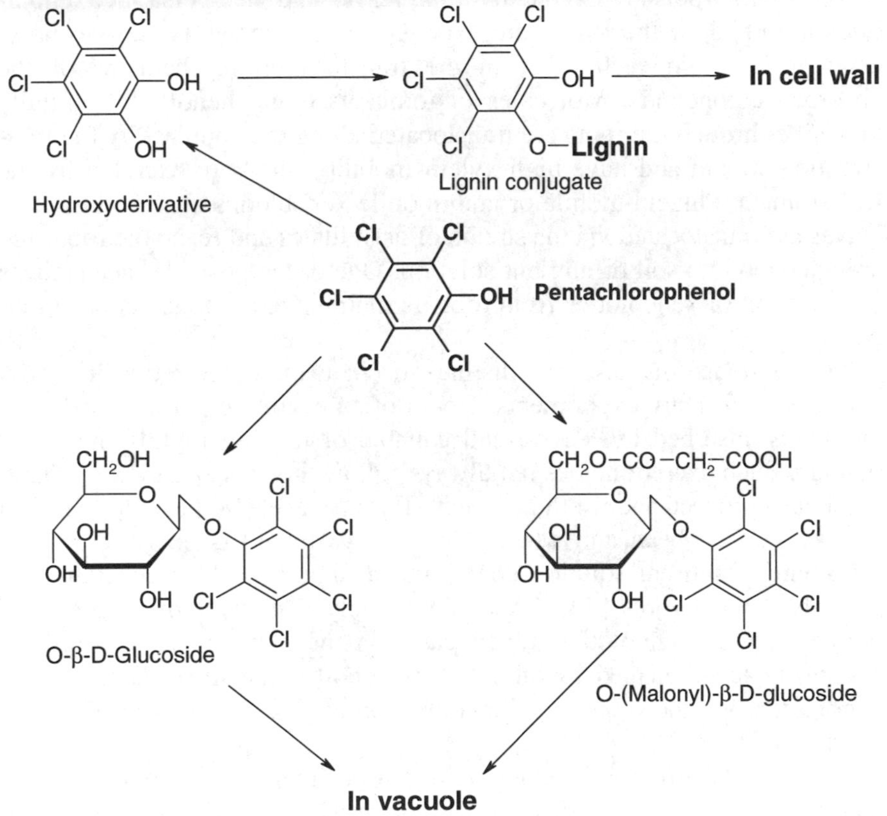

Fig. 3.4. Pentachlorophenol transformation for deposition in vacuoles and the cell wall

The mechanisms of excretion, degradation, and conjugation of toxic compounds are discussed below.

3.2 Excretion

The term "excretion" in plants implies the partial release of environmental contaminants absorbed by the plant in unchanged form through the leaves or the root system [278, 544]. It therefore differs from excretion in mammals and other animals, in which transformed as well as untransformed substances are eliminated from the body. Environmental pollutants absorbed by the roots are excreted via the leaves and vice versa, i.e. xenobiotics absorbed by the leaves are excreted via the roots. These two processes differ from each other in the translocation mechanisms of the absorbed compounds. Molecules of toxicants (e.g. phenol) that initially penetrate through the roots are translocated along the apoplast by the transpiration stream and have high xylem mobility; they are excreted by the leaf stomata. Phloem-mobile or ambimobile xenobiotics absorbed through leaves are translocated via the stream of assimilates and reach the roots and are excreted into soil or nutrient solution. One of the possible mechanisms of excretion of xenobiotics from roots is their excretion together with mucus.

The existence of these two distinct mechanisms of excretion has been confirmed by many experiments. Roots often excrete environmental contaminants absorbed by leaves, and capable of moving rapidly along the phloem. Such excretion does not always follow the concentration gradient; it can also proceed against the gradient. For instance, [14C] alachlor applied to leaves of soybean and wheat (*Triticum aestivum*) is excreted via the roots into a nutrient solution containing a higher alachlor concentration than that in the roots [73]. These data imply that the root excretion of phloem-mobile xenobiotics, translocated in the plant by the assimilates stream, is accomplished by an active transport mechanism. Excretion of xenobiotics via the roots is a functional process, characteristic of higher plants.

Besides phloem-mobile compounds, sometimes ambimobile environmental contaminants absorbed by the leaf surface are excreted in untransformed form via the roots. Root excretion is especially characteristic for the phenoxyacetic acids (2,4-D, 2,4,5-T, etc.), dicamba, picloram and other systemic herbicides [211, 312, 444]. The root system more actively excretes absorbed xenobiotics than the leaves. For instance, the roots of

Ampelamus albidus excrete approximately 37% of the total amount absorbed (by the plant leaves) of 2,4-D over 8 days [113].

The closer the contaminant-absorbing leaf is to the roots the higher is the rate of excretion by the roots [444]. The intensity of excretion also increases with an increase in herbicide concentration applied to the leaf. Excretion often proceeds via the roots of both herbicide-sensitive and herbicide-resistant plants, and there is no clear relation between excretion and plant resistance to herbicide [113].

The root system also excretes toxic compounds absorbed by the roots or by the stem. For instance, cotton seedlings excrete backwards through the roots about 25–30% of the herbicide bioxone, which was absorbed by the roots from herbicide-containing nutrient solution [258]. Such behavior is observed after transfer of seedlings to herbicide-free nutrient solution for 2 days. In another example, 15% of the total amount of the hydrazide of maleic acid injected over 30 days into the wood of saplings of rock maple (*Acer saccharinum*) and western plane (*Platanus occidentalis*) was excreted from the roots in untransformed form [119].

In the examples presented above most of the absorbed contaminants were excreted without transformation by the root systems; generally the amount of excreted xenobiotic varied between 0.1–2%. Nevertheless, the phenomenon of root excretion must be taken into account when considering the treatment of plant surfaces with different pesticides, since excessive excretion of xenobiotics in unchanged form could become a source of serious chemical contamination of soil and groundwater.

Environmental pollutants absorbed by the roots can also be excreted via the leaves, although this excretion is rare as compared to root excretion. A good example is the excretion of phenol by the leaves of bulrush (*Scirpus lacustris* L.) plants kept on phenol solution [449]. In this case the excretion proceeds so rapidly that after 90 min phenol can be detected in the air near the leaves and after several hours the phenol in the air can be detected even by smell. Another example shows how leaves of tobacco and radish immersed with their petioles in a solution of 1,2-dibromoethane absorb and then rapidly excrete the xenobiotic into the atmosphere [249]. The inference from these and other analogous data is that plants can excrete halogen derivatives of hydrocarbons absorbed from the soil or groundwater and gradually dilute them into the air. The occurrence of this process is further supported by laboratory and field experiments on poplar hybrids removing TCE from artificially contaminated (260 mg/l) water and soil [264]. Less than 10% of the total TCE assimilated by the plant was evaporated via the poplar leaves without transformation and the rest was metabolized. This example indicates once more that excretion of absorbed xenobiotics proceeds at high concentrations. In this case, the organic contaminant penetrates rapidly along the xylem and is not transformed before it is evaporated through the stomata together with water. In this way only a relatively insignificant proportion of the absorbed xenobiotic can be excreted.

Considering the process of excretion as a method of phytoremediation, it should be mentioned that the ability of higher plants to assimilate environmental contaminants via their root systems and excrete them in untransformed form via the leaves could be used for the purification of soils with high concentrations of xenobiotics.

3.3 Transformation

3.3.1 Conjugation with endogenous compounds

Detoxification via conjugation is one of the defense mechanisms of higher plants. The toxicity of conjugates compared to parent compounds is significantly decreased because of binding with non-toxic cellular compounds. Conjugates are kept in a cell for a certain period of time without causing visible pathological deviation from cell homeostasis.

Conjugation of environmental contaminants (xenobiotics) of different structures with cellular metabolites represents a special detoxification mechanism, particularly in those cases where the concentration of the contaminants that have penetrated into the plant cell exceeds the plant's potential for deep transformation (degradation). In these cases, the main part of the toxicant undergoes conjugation and only a small part is metabolized (5 to 30%, the transformation above 10% is achieved in the case of linear low molecular mass hydrocarbons) to common plant metabolites, or to CO_2 and water.

Conjugate formation also gives the plant cell more time for the induction of enzymes responsible for contaminant degradation (mainly oxidative). This may enable the induction of enzymes by very hydrophobic or structurally complicated contaminants. In the great majority of cases, relatively quickly after the termination of plant incubation with the xenobiotic (and possibly some delay thereafter depending on xenobiotic nature, plant type, and duration of exposure) conjugates are no longer found in plant cells. This indicates that conjugates eventually release their toxic part unchanged, and further undergoes mineralization by plant cell enzymes [84].

As a result of conjugation, xenobiotics are coupled with endogenic sugars, amino acids and peptides through existing hydrophilic functional groups, or such groups acquired after initial introduction of the xenobiotic into the plant cell. Conjugation reactions and enzymes catalyzing these processes (transferases) are described in this Chapter.

Conjugation with carbohydrates

Among the known mechanisms of conjugation, glycosylation is one of the most widespread in higher plants. Xenobiotics with hydroxyl, carboxyl, amine, or other functional groups as a constituent part of their molecules, are directly subjected to glycosylation. Often xenobiotics are exposed to minor initial transformations after penetrating into the plant cell, in which such groups are introduced into their molecules, significantly increasing their reactivity.

Alcohols and phenols often undergo glycosylation in plants. This is illustrated by many examples indicating the formation of β-glucosides. Thus, formation of ethyl-β-glucoside was observed during cultivation of seedlings of mung bean in an ethanol-containing zone [338].

The metabolic fate of $[1-6-^{14}C]$ 3,4-dichloroaniline (DCA) was investigated in thale cress *(Arabidopsis thaliana)* root cultures and soybean plants after a 48 h treatment via the roots. DCA is rapidly taken up by both species and predominantly metabolized to N-malonyl-DCA in soybean and N-glucosyl-DCA in *Arabidopsis*. Synthesis is observed in roots and the respective conjugates are largely exported into the culture medium, with only a minority retained within the plant tissue. Once conjugated, the roots of both species slowly take up the DCA metabolites from the medium. The difference in the routes of DCA detoxification in two plants could be explained partly by the relative activities of the conjugating enzymes, soybean having high DCA-N-malonyltransferase activity, and *Arabidopsis* a high DCA-N-glucosyltransferase activity [291].

DCA is rapidly conjugated by glucosylation in *Arabidopsis* root cultures by transformation into N-β-D-glucopyranosyl-DCA and excretion into the medium. The enzyme N-glucosyltransferase is responsible for this transformation [318].

As a result of injection into apple *(Malus sylvestris)*, geraniol is subjected to glucosylation to geranyl-β-D-glucoside [533]. Glucosylation of a foreign alcohol hydroxyl group may lead to glucoside formation from saligenin. In broad bean seedlings *o*-hydroxybenzyl-β-glucoside was formed in tissues with introduced saligenin [388]. In this case not the phenolic hydroxyl but exclusively the alcohol hydroxyl group is glucosylated. The study of saligenin transformation by a suspension culture of datura *(Datura innoxia)* demonstrated that the main metabolite is glucoside with alcohol hydroxyl groups and to a far lesser extent glucoside with phenol hydroxyl groups [490]. The herbicide N-hydroxymethyl dimethoate is also glucosylated via a free primary alcohol hydroxyl group [185].

Pentachlorophenol is glucosylated in wheat and soybean plants via combination with malonic acid and transformed to β-D-glucoside and O-malonyl-β-D-glucoside conjugates (see Fig. 3.4) [435]. Pridham [388] showed that mono-, di-, and triatomic phenols are easily (enzymatically) converted into the corresponding β-monoglucosides in experiments with broad bean seedlings.

In some cases when phenols are glycosylated, the existence of di- and triglycosides has been demonstrated. For instance, diglycoside (gentiobioside) and triglycosides are formed from exogenous hydroquinone in wheat embryos [217].

Often, hydroxyl derivatives are produced among the primary products of transformations of the xenobiotics in plant tissues, and are further subjected to rapid glycosylation. Thus, a conjugate is formed from the oxidation product of the systemic fungicide etirimol. The aliphatic side chain (butyl group) of this herbicide is oxidized and the alcoholic hydroxyl formed is glycosylated in leaves of barley [217]. The herbicide diphenamid is oxidized, i.e., the N-methyl group is hydroxylated in pepper seedlings [233] and in callus tissue of tobacco [62].

The carboxyl group of xenobiotics is one of the most reactive functional groups, and this is why it often undergoes glycosylation in plants. For instance, the formation of esters with glucose is characteristic of phenoxyacetic acids. In root callus tissues of rice *(Oryza sativa)* grown in a liquid nutrient medium containing [^{14}C] 2,4-D, the glucosyl ester of 2,4-D is the main product isolated. However, no amino acid conjugates of 2,4-D have been found that were identified in callus tissues of several other plants. Thus, the main pathway of 2,4-D metabolism in rice root callus tissue is the formation of esters with glucose, although other pathways may operate in other plant species [166].

2,4-D glucose esters occur widely and in large amounts in herbicide-resistant wild wheat plants *(Triticum dicoccum),* timothy *(Phleum pratense),* and kidney bean [80].

O-β-D-Glucoside of 2,4-D

Besides the carboxyl group, other acidic groups are also subjected to glycosylation in plants. Thus, the plant growth regulator ethephon is glycosylated by the formation of β-D-glucopyranoside-1-(2-chloroethyl)-phosphonate in bark cuts of hevea (*Hevea brasiliensis*) [20].

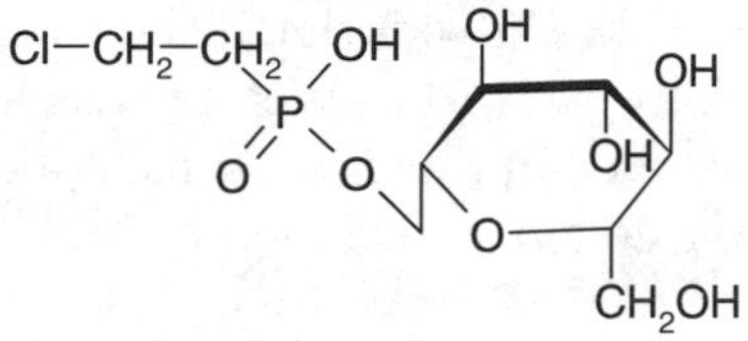

O-β-D-Glucoside of 1-(2-chloroethyl)phosphonate

As it has been indicated, in some cases, sugars other than glucose also participate in an esterification reaction with the carboxyl group of xenobiotics. For instance, arabinose becomes esterified with nicotinic acid in culture suspensions of parsley *(Petroselinum sativum)* [304].

Glycosylation via blocking of the free amino groups of xenobiotics is another widespread mechanism of conjugation, since xenobiotics of different structures often contain such groups. Thus, 3-amino-2,5-dichlorobenzoic acid is further transformed into the corresponding N- and O-glucosides after its initial transformation to the glucose ester (Fig. 3.5) in roots, shoots, and hypocotyls of *Setaria* sp. [175].

In studies on the glycosylation of synthetic cytokinin analogues in rootless seedlings of radish, the conversion of amides of 4-(purine-6-yl-amino)-butyric acid, 6-(3,4-dimethoxybenzyl-amino)-purine, and 6-benzylamino-purine into the corresponding 7-glucopyranosides has been demonstrated,

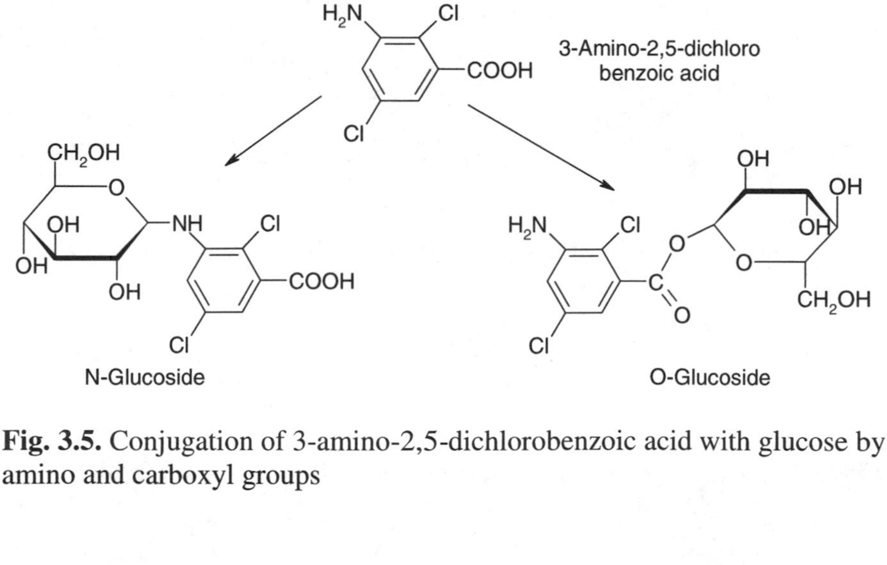

Fig. 3.5. Conjugation of 3-amino-2,5-dichlorobenzoic acid with glucose by amino and carboxyl groups

but adenine and methylaminopurine are not glycosylated under the same conditions [306]. Ribosides are formed as a result of the absorption of 6-benzylaminopurine by the roots of kidney bean seedlings [394]. 4-chloroaniline and 3,4-DCA are glycosylated, and malonic conjugates are formed in wheat plants and culture suspensions of wheat and soybean [531]. The herbicide metribuzin is first glycosylated and subsequently conjugated with malonic acid in tomato [177]. However, the results of other studies on tomato biotypes with high, medium, and low sensitivities to metribuzin indicate that N-glucoside is the dominant metabolite [469].

Conjugation with amino acids

Amino acids, being multifunctional secondary metabolites, participate in many processes vitally important for plant cells. Recently it was shown that in addition to all their previously known attributes of metabolic importance, plants effectively participate in the detoxification (conjugation) of a broad spectrum of organic contaminants.

Conjugation with amino acids is a widespread reaction of the carboxyl group of xenobiotics in plants. A study of 2,4-D metabolism in soybean species has demonstrated that the glycoside conjugate of 4-oxy-2,5-dichlorophenoxyacetic acid is the primary metabolite in resistant species, but that in sensitive species forms conjugates with amino acids [528]. 2,4-D forms conjugates with glutamic and aspartic acids (Fig. 3.6) in callus and differenti-

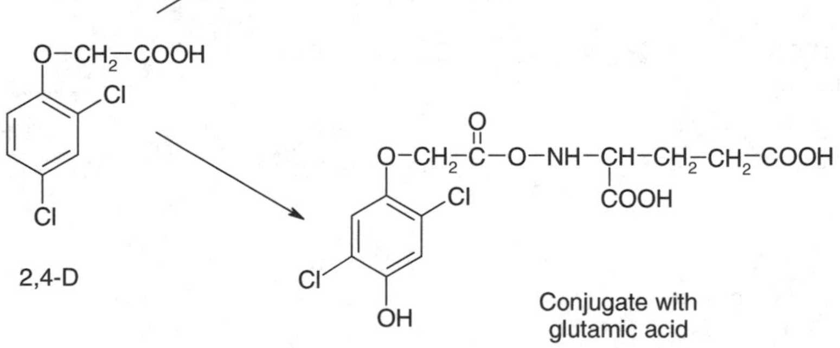

Fig. 3.6. Conjugation of 2,4-D hydroxy-derivative with amino acids

ated root tissues of soybean [99], in tissue cultures of maize endosperm, and in the medullar parenchyma of tobacco, carrot, and sunflower [166].

ε-(2,4-dichlorophenoxyacetyl)-L-Lys (2,4-D-Lys) is a widely distributed conjugate [105]. The uptake of 2,4-D-Lys by broad bean (*Vicia faba*) leaf discs is mediated by an active carrier system ($K_{m1} = 0.2$ mM; $V_{max1} = 2.4$ nmol cm^{-2} h^{-1} at pH 5.0) and complemented by an important diffusive component. Among the compounds tested, i.e., neutral, basic, and acidic amino acids, auxin, glutathione, sugars and also various herbicides and fungicides with large-size lipophilic groups, the aromatic amino acids evidently compete with 2,4-D-Lys as a product of conjugation. The conjugate accumulates in the vein network of the leaves, is exported towards the growing organs, and exhibits a distribution pattern different from that of the unconjugated herbicide. The chlorinated conjugate 2,4-D-Lys has been chosen for several studies for the following reasons: this compound is one of the largest molecules previously tested and one of the most efficient inhibitors of the amino acid transport system in leaf tissues of broad bean [81], undoubtedly playing an important role in amino acid metabolism.

Conjugation with peptides, proteins, lignin and hemicellulose

The variety of peptides and proteins in plant cells, with various functional groups, and their ability to couple with xenobiotics of different structures, creates several favorable opportunities for xenobiotics conjugation.

One of the most important pathways for maintaining the health of plants via detoxification is the conjugation of xenobiotics with the tripeptide – reduced glutathione (GSH) (γ-Glu–Cys–Gly). This detoxification pathway is most characteristic for symmetric triazines, chloroacetamides, and other halogen-containing compounds. A study on atrazine transformation in 53 herbaceous plant species *(Festucae, Avenae, Triticeae, Paniceae, Andropogonae, Eragrosteae, Chlorideae)* revealed that the herbicide formed conjugates with glutathione in all plants tested [256]. An analysis of the atrazine transformation products formed in herbicide-resistant and herbicide-sensitive herbs revealed that in the resistant plants, big bluestem *(Andropogon gerardii* Vitman*)* and switch-grass *(Panicum virgatum),* the major metabolite is the atrazine conjugate with glutathione. In the sensitive plants Indian grass *(Sorghastrum nutans)* and sideoats grama *(Bouteloua curti-pendula* (Michx.) Torr.), mainly N-de-ethylation products are formed [523]. Studies on the metabolism of atrazine in culture suspensions of wheat and potato *(Solanum tuberosum)* have indicated that the herbicide is transformed by N-de-ethylation in wheat cells, and by conjugation with glutathione in potato cells. The enzyme glutathione S-transferase (GST), capable of using atrazine as a substrate, was also identified from potato cells [143]. Conjugation with glutathione is characteristic for chloroacetamide herbicides [298]. For example, acetochlor forms conjugates with glutathione in seedlings of maize, morning glory *(Ipomoea purpurea)*, field bindweed *(Convolvulus arvensis)*, cocklebur *(Xanthium pensylvanicum)*, and velvetleaf *(Abutilon theophrasti)* [42]. Applied to coleoptiles of maize seedlings, alachlor and metolachlor form glutathione conjugates. The GST isolated from maize seedlings, which catalyzes xenobiotic conjugation with glutathione, has a threefold higher activity with alachlor as a substrate than with metolachlor [361]. Active transformation of pretilachlor into its glutathione conjugate was observed in rice seedlings [212]. It was also suggested that compounds like benzyl chloride and propachlor form conjugates with glutathione by reactions catalyzed both enzymatically and non-enzymatically (Fig. 3.7) [212].

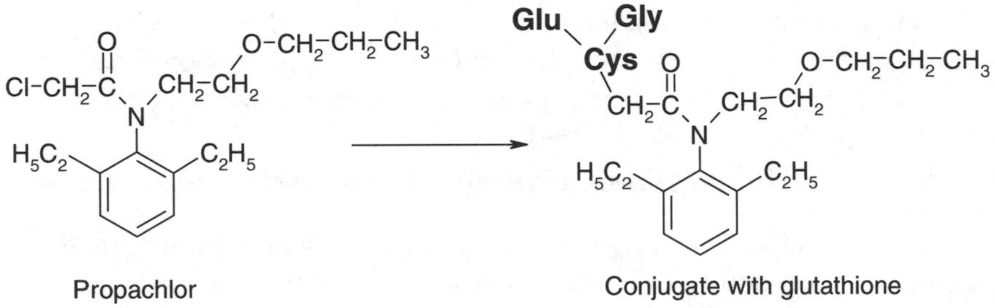

Fig. 3.7. Conjugation of propachlor with reduced glutathione

Another tripeptide, homoglutathione (differing from glutathione by containing β-alanine instead of glycine), may also participate in conjugation reactions with xenobiotics in plants. The formation of homoglutathione conjugates is characteristic mainly for soybean. Thus, the herbicide propachlor forms a conjugate with homoglutathione in soybean seedlings [289]. The same transformation occurs during chlorimuron-ethyl and thifensulfuron-methyl metabolism [46, 47]. Homoglutathionic conjugates of acetochlor are formed also in other plants, particularly in soybean, mung bean, and alfalfa [42].

Glutathione and homoglutathione conjugates of xenobiotics with a hydroxyl group are formed *in vivo*. Thus, acifluorfen, a derivative of diphenyl ether, is cleaved into 2-nitro-5-oxybenzoic acid in soybean seedlings, which couples with homoglutathione via its hydroxyl group [176].

Another mechanism characteristic of the binding of xenobiotics to glutathione and homoglutathione is the reaction with alkylthio groups. The S-ethyldipropyl thiocarbamate conjugates with glutathione via an ethyl group in maize seedlings [66, 297]. It is supposed that in this particular case the herbicide initially is oxidized into the corresponding sulfoxide and is then conjugated with glutathione. The latter process is catalyzed by GST. Metribuzin binds with homoglutathione via a methylthio group in soybean [178]. 3,4-benzopyrene is oxidized by conjugation with glutathione in microsomes from parsley cell suspensions, soybean and primary leaves of pea seedlings [503].

Phenol (oxybenzene) is not glycosylated in intact plants of maize, pea, and pumpkin (*Cucurbita pepo*). A study of [1– 6-^{14}C] phenol metabolism in sterile seedlings of these plants has demonstrated that phenol forms conjugates with low-molecular-mass peptides in plants [83, 515]. Other monophenols also form conjugates with peptides in plants, namely α-naphthol in maize, pea, and pumpkin seedlings [509, 513]; *o*-nitrophenol in pea seedlings [509, 513]; and a hydroxyl derivative of 2,4-D in maize, pumpkin, and pea seedlings [16]. Phenols are covalently bound to peptides via hydroxyl groups. The amino acid composition of peptides participating in the conjugation of phenols varies. In plants treated with phenol, the low-molecular-mass peptide concentration increases [513]. In some plants, conjugation with low-molecular-mass peptides seems to be an important detoxification pathway for monophenols. Phenoxyacetic acids introduced into plant tissues form peptide conjugates. In sterile seedlings of maize and snap bean, phenoxyacetic acid and 2,4-D form conjugates with low-molecular-mass peptides. In vine, conjugates of these acids with peptides are also formed [262, 341]. As a result of hydrolysis of phenoxyacetic acid and 2,4-D, peptide conjugates with 6 to 10 amino acids are formed [138, 261, 262]. In cereals, peptides/proteins participating in conjugation with phenoxyacetic acid contain 2 to 220 amino acid residues [79].

Quite often organic contaminants are coupled with cell biopolymers participating in the formation of cell wall structure. A good example is the metabolism of [^{14}C] TNT absorbed by roots of kidney bean. TNT conjugates with lignin (20%), hemicellulose (14%), and pectin (5%) [451].

Lignin is a phenolic, structurally nonrepeating macromolecule, which is active in conjugation reactions, and often plays the role of a carrier of xenobiotics and their primary transformants [428]. Such compounds are incorporated into the lignin structure by being covalently coupled with the biopolymer. It has been shown that tautomeric forms of the lignin monomer coniferyl alcohol (quinone-methyl) couple xenobiotics with amino and hydroxyl groups, as for instance 3-chloroaniline (Fig. 3.8) [429].

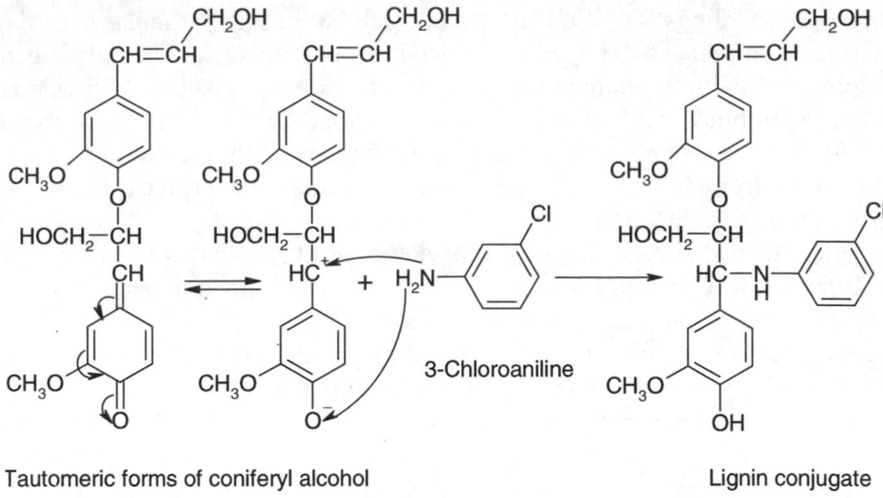

Fig. 3.8. Conjugation of 3-chloroaniline with coniferyl alcohol

An analogous picture is observed in the case of pentachlorophenol. Coniferyl alcohol easily conjugates with 1,2-dihydroxy-3,4,5,6-tetrachlorobenzene, intermediate of pentachlorophenol hydroxylation (Fig. 3.9) [426].

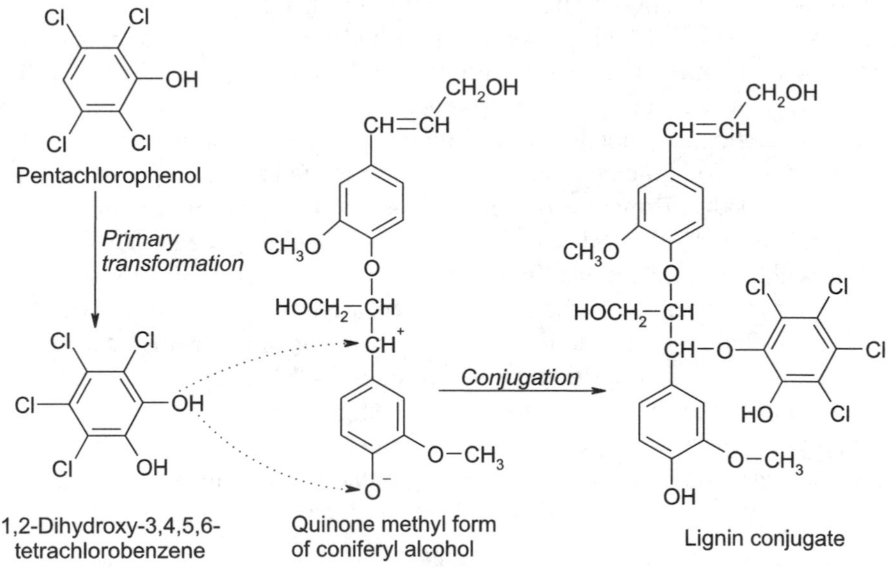

Fig. 3.9. Conjugation of pentachlorophenol with coniferyl alcohol

Borken and Harms [39], investigating the distribution of [^{14}C] 4-nonylphenol (a surfactant) in suspension cultures of 12 plant species, have found that in 7 of the cultures most of the xenobiotic is conjugated with lignin. Lignification of triazols was also demonstrated in sunflower [69]. Lignin covalently binds with DDT residues in Canadian waterweed (*Elodea canadensis*) [184].

The metabolism of [^{14}C] phenoxyacetic acid and the formation of associated bound residues in soybean leaves and stem has also been reported [60]. Lignin is not the only biopolymer involved in binding with xenobiotics. In the leaves, xylan and lignin are the preferred compounds for the binding of labelled carbon atom, whereas in the stems pectin and lignin are the main components binding the radioactive label [^{14}C] of this herbicide.

Xylans (hemicelluloses), being widely present in plant tissues and possessing many free carboxyl groups, actively participate in the conjugation with amino or hydroxyl groups of xenobiotic molecules. A good example of such a combination is the conjugation of ADNTs, primary products of TNT reduction, with hemicellulose (Fig. 3.10) in the roots of hybrid willow (*Salix* sp.), and Norway spruce, trees used in dendroremediation of soils polluted by TNT [440].

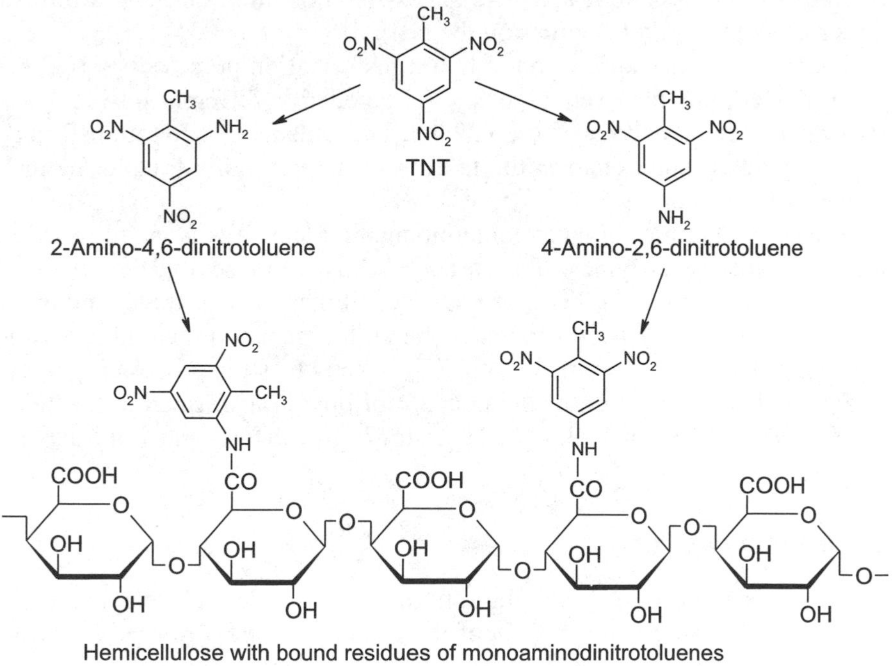
TNT
2-Amino-4,6-dinitrotoluene
4-Amino-2,6-dinitrotoluene
Hemicellulose with bound residues of monoaminodinitrotoluenes

Fig. 3.10. Conjugation of TNT metabolites with hemicellulose

In summary, according to the existing experimental information, environmental contaminants may be conjugated directly with biopolymers, or coupled with monomers and undergo copolymerization to form a modified biopolymer.

3.3.2 Degradation pathways

One the most desirable ecological features of plants is their potential to carry out deep degradation (oxidation) of environmental organic contaminants. Such transformations lead to the partial decomposition of the carbon skeletons of toxic molecules to regular cell metabolites, or to mineralization to CO_2 and further participation of carbon atoms in characteristic natural cycles (plant cell processes). Depending on the plant species, the nature of the contaminant and its concentration, a relatively small proportion of the environmental contaminant penetrating into the plant cell undergoes deep oxidation. At high xenobiotic concentrations most of the contaminant, that finds its way into the cell, is conjugated and deposited inside the cell. Ultimately it is released from the sites of deposition and degraded. Conjugation thus allows the plant to defer the possible energetically costly degradation process until it is in a stress-free condition and can afford to expend the necessary resources on the task.

The mass balance approach to phytotransformation processes is preferably followed, in which xenobiotics [^{14}C]-labelled in different positions are used as tracers. Analysis of the labelled and unlabelled phytotransformation products usually enables the tracing of all metabolic transformations of each carbon atom.

The degradation of organic contaminants in plant cells is mainly carried out by oxidative enzymes. This process consists of several consecutive stages and ends by the release of CO_2 (sometimes only in trace amounts). Since carbon dioxide is considered to be an inorganic compound, the conversion of the organic pollutant into CO_2 is known as the process of mineralization. The verification of the concept of phytotransformation has been confirmed at different scales: in laboratory, greenhouse and field experiments.

Hydrolytic cleavage

In most cases the ester bonds in xenobiotic molecules are enzymatically cleaved. For example, up to 95% of the triclopyr esters absorbed by resistant wheat, tolerant barley, and sensitive common chickweed plants are hydrolyzed in the 3 days following treatment, and conjugated with glucose and aspartic acid [308].

Three derivatives of sulfonylurea are metabolized with different rates in soybean plants, indicating that cleavage rates of the ester bonds are different in the three derivatives [46, 47]. Thifensulfuron-methyl is rapidly hydrolyzed into the corresponding thifensulfuronic acid, while the half-life of the thifensulfuric acid methyl ester in plant tissue is 4–6 h. Another ester, chlorimuron-ethyl, is also de-esterified, but more slowly, while conjugation with homoglutathione prevails. Metsulfuron-methyl, the third ester, does not undergo de-esterification under the same conditions in the same soybean seedlings.

Xenobiotics with ether bonds are transformed mainly at the ether site. When ether bonds are lacking, other easily oxidized side groups of the diphenyl ether system are transformed, and only if the latter are absent, cleavage of ether bonds occurs in wheat seedlings [251, 492], oats [251], oat cell suspension culture [457], and ryegrass [456]. Difenopenten-ethyl is de-etherified in soybean and wheat seedlings [458]. The highly selective diphenyl ether herbicide AKH-7088 is metabolized in soybean by complete oxidation of the side chain [279]. However, the acifluorfen molecule is cleaved at the ether bond in soybean to form the corresponding phenols, which are then conjugated directly with glucose (and subsequently with malonic acid) and homoglutathione (Fig. 3.11) [176]. Similarly, fluorodifen, nitrofen [461], and other diphenyl ethers are cleaved into the corresponding phenols. All these phenols are directly transformed into corresponding conjugates.

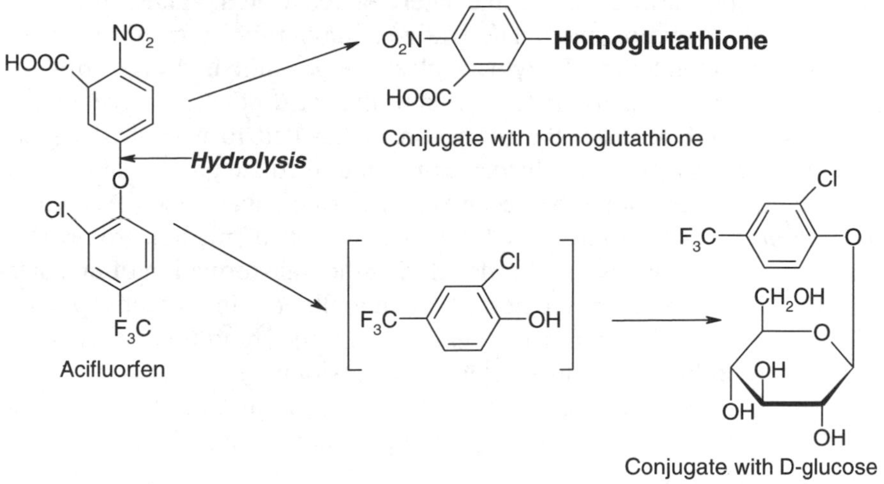

Fig. 3.11. Hydrolysis and conjugation of acifluorfen

Hydroxylation

The introduction of different hydrophilic groups is required to increase the reactivity of xenobiotics for their further cellular transformations. Introduction of a hydroxyl group into a xenobiotic molecule increases its polarity and hydrophilicity. In a number of cases, hydroxylation is the primary detoxification reaction, followed by profound oxidation and conjugation [428]. Studies of the metabolites of xenobiotic alkanes and N-alkyl derivatives indicate that oxidative degradation of these molecules often begins with hydroxylation of the alkyl groups. Although it is not always possible to isolate and identify the corresponding hydroxy derivatives, the products of their further metabolism provide information on the chemical structures of the intermediates.

Low-molecular-mass $(C_1–C_5)$ $[^{14}C]$-labelled alkanes absorbed by leaves are subjected to oxidative transformation to $^{14}CO_2$ [128–131]. On the basis of identified intermediate products, it was concluded that these hydrocarbons are oxidized monoterminally, with formation of the corresponding primary alcohols, followed by oxidation to carboxyl acids (see the section "Deep oxidation").

The hydroxylation of N-alkyl groups is a characteristic reaction in the transformation of urea-based herbicides in plants. Fast oxidation of the hydroxy-alkyl groups formed is accompanied by hydroxylation, generating a dealkylated product. N-Dealkylation is the primary metabolic transformation pathway of N-methylphenyl urea herbicides. In some cases the hydroxyl groups formed are immediately glucosylated. Thus, the β-D-glucoside of the hydroxymethyl derivative of monuron is formed from $[^{14}C]$ monuron in cotton leaves. Enzymatic cleavage or acidic hydrolysis of these glycosidic bonds leads to the formation of the corresponding demethylated products. Simultaneously, the formation of labelled formaldehyde is observed [174]. An analogous glucoside of an intermediate product of a diuron hydroxymethyl derivative has been isolated from sugar cane *(Saccharum officinarum)* [314]. Products of the hydroxylation of methyl groups (hydroxymethyl derivatives) are formed during the transformation of urea herbicides in plants: buturon in wheat [216], monolinuron in spinach *(Spinacia oleracea)* [445], tebuthiuron in sugar cane [316], chlorotoluron in wheat [202]. Chlorotoluron is hydroxylated in two positions (Fig. 3.12):

I position – hydroxylation of the N-methyl group leads to the formation of an unstable N-hydroxymethyl group, which is degraded, and finally demethylation takes place;

II position – hydroxylation of the methyl group bound to the aromatic ring leads to the formation of stable products (in contrast to the N-hydroxymethyl group, the C-hydroxymethyl group is stable) [202].

Both products are formed in herbicide-resistant and sensitive varieties of wheat [65], but not in equal amounts. The dominant metabolite is formed by N-demethylation (5.8%); the C-hydroxymethyl derivative is a minor component (1.4%). Analogously, 2-*sec*-butylphenol-N-methylcarbamate is transformed by C-hydroxylation and N-demethylation. This herbicide is absorbed by rice and undergoes hydroxylation by *sec*-butyl as well as by N-methyl groups [362].

Symmetric triazines are subjected to N-dealkylation in plants. In the case of triazine herbicides, N-dealkylation proceeds by hydroxylation of a side chain (alkyl group), but the corresponding hydroxy derivatives have not been identified [384, 523, 529]. However, despite the fact that the hydroxy derivative of atrazine is not found in culture suspension of potato and wheat, the product of hydroxylation of another symmetric triazine, terbutryn, has been isolated, and appears to be the major metabolite [143]. Moreover, it was found that [2-^{14}C] terbacil, once absorbed by alfalfa, is hydroxylated via the methyl group [402].

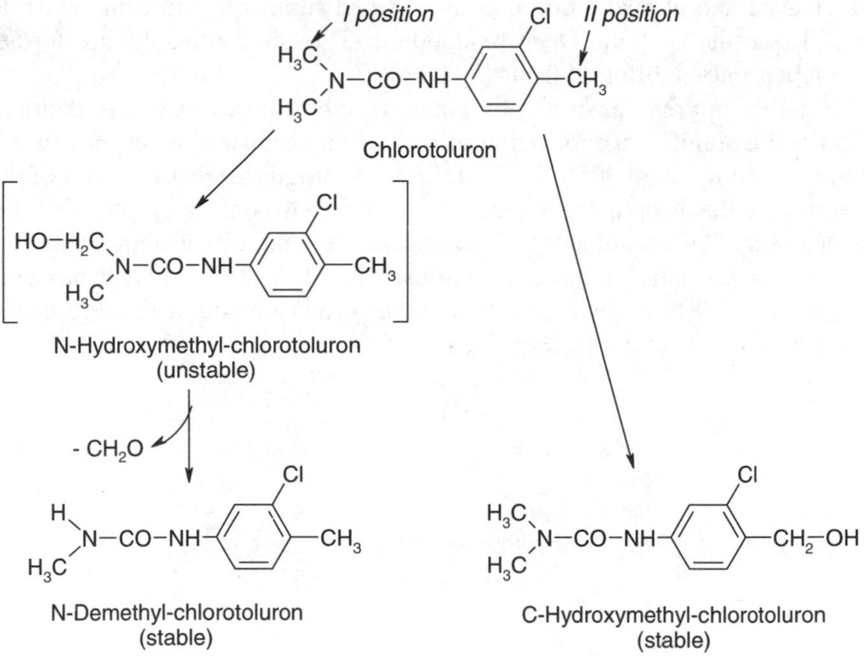

Fig. 3.12. Hydroxylation of chlorotoluron in two positions

Hydroxylation of the methylene group of xenobiotics has also been reported several times. Thus, carbofuran is hydroxylated at the C_3-atom in barley, maize [371], and strawberry [15].

The transformation of [^{14}C] cyclohexane in plants indicates that the ring of this hydrocarbon is cleaved, and aliphatic products are formed. The first step of cyclohexane transformation in plants is its hydroxylation into cyclohexanol (Fig. 3.13) [508].

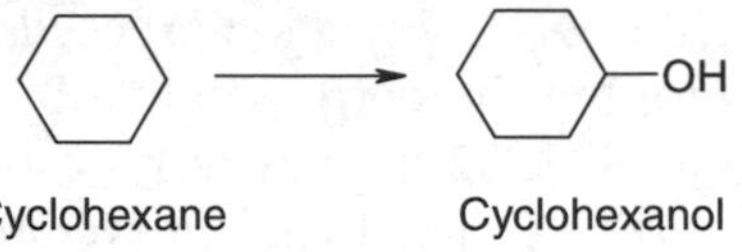

Cyclohexane   Cyclohexanol

Fig. 3.13. Hydroxylation of cyclohexane

Similarly, the first step in the metabolism of aromatic hydrocarbons in plants is formation of hydroxy-derivative intermediates. [1–6-^{14}C] benzene is cleaved and aliphatic products are formed (muconic and fumaric acids) [133] (see the section "Deep oxidation"). The same products are formed from benzene in different fruits [134].

3,4-Benzopyrene absorbed by plants is subjected to oxidative degradation and a significant proportion of its carbon atoms are incorporated into aliphatic compounds [218, 219, 503]. The analogous transformation of this xenobiotic has been determined in cell culture suspensions [108, 109]. For PAHs such as naphthalene, benzanthracene and dibenzanthracene, the same transformation pathways are observed [108, 110, 111]. It has been suggested [108] that hydroxylation is the primary reaction in the transformation of PAHs in plants (Fig. 3.14–3.15).

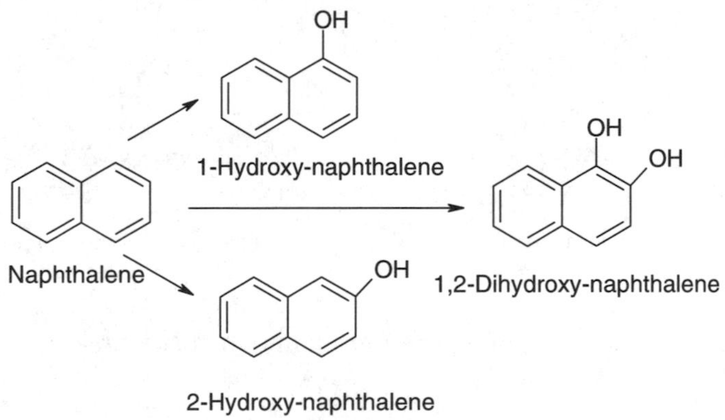

OH

1-Hydroxy-naphthalene

OH
OH

Naphthalene   1,2-Dihydroxy-naphthalene

OH

2-Hydroxy-naphthalene

Fig. 3.14. Hydroxylation of naphthalene

Hydroxylation of the aromatic ring is an important step in the transformation of phenoxyacetic acid in plants. The introduced hydroxyl group is often subjected to glycosylation. Phenoxyacetic acid is hydroxylated most often at position 4 in the aromatic ring. The hydroxylase activity increases 16-fold during the formation of the hydroxylated metabolite of phenoxyacetic acid (4-hydroxyphenoxyacetic acid) in oat seed embryos [244]. Phenoxyacetic acid, halogenated in the aromatic ring, is hydroxylated at non-substituted carbon atoms of the benzene ring. However, hydroxylation of 2,4-D often occurs at position 4 and the chlorine atom moves to the positions 5 or 3. For example, identification of hydroxylated 2,4-D compounds in different plants such as wild buckwheat *(Polygonum convolvulus)*, leafy spurge *(Euphorbia esula)*, yellow foxtail *(Setaria glauca)*, wild oat *(Avena fatua)*, wild mustard, perennial sowthistle *(Sonchus arvensis)* and kochia *(Kochia scoparia)* has revealed that 2,5-dichloro-4-hydroxyphenoxyacetic acid is the dominant metabolite in all plants studied [168]. Investigation of the transformation of 2,4-D in herbicide-sensitive and herbicide-resistant soybean species has demonstrated that the 4-hydroxy derivative of 2,4-D is mostly formed in resistant species, and exclusively in the form of a glycoside [528].

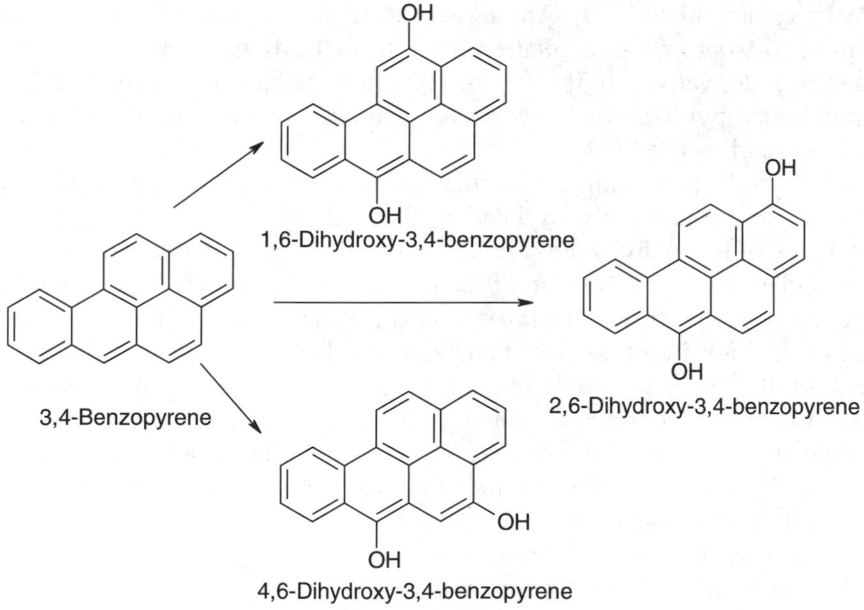

1,6-Dihydroxy-3,4-benzopyrene

2,6-Dihydroxy-3,4-benzopyrene

3,4-Benzopyrene

4,6-Dihydroxy-3,4-benzopyrene

Fig. 3.15. Hydroxylation of 3,4-benzopyrene

The herbicide diclofop [457] and its methyl ether (diclofop-methyl) [492] are hydroxylated in plants in a similar way, although the product formed by glycosylation of the carboxyl group in some plants species is the dominant metabolite [251]. The tolerance of resistant and sensitive ryegrass (*Lolium rigidum*) biotypes to diclofop does not depend on pesticide metabolites, since a considerable amount of the phytotoxic diclofop as well as its conjugates and hydroxylated ring derivatives are formed in the stems and roots of both biotypes [456]. The enzyme catalyzing the transformation of diclofop into 2-β-(2,5-dichloro-4-hydroxyphenoxy)phenoxy-propionic acid has been successfully isolated and purified from etiolated wheat seedlings [331].

Various plants hydroxylate benzoic acid and its derivatives. Benzoic acid is hydroxylated simultaneously at the *o*- and *p*-positions, and sometimes both hydroxy acids occur in tissues. It has also been shown that dicamba is hydroxylated at position 5 and this product is the main one among the herbicide metabolites in many plants [74, 410].

Isopropyl carbanilate and isopropyl *m*-chlorocarbanilate are also subjected to hydroxylation of the aromatic ring at different positions with subsequent conjugation of the hydroxy derivatives with glucose. Isopropyl carbanilate in alfalfa is transformed mainly into isopropyl-4-hydroxycarbanilate [484]. Among wheat, sugar beet and alfalfa plants exposed to isopropyl carbanilate, only wheat forms the 4-hydroxy and 2-hydroxy derivatives [63]. Isopropyl-3-chlorocarbanilate is hydroxylated into isopropyl-3-chloro-2-hydroxycarbanilate or isopropyl-3-chloro-4-hydroxycarbanilate [484].

The products resulting from the hydroxylation of aromatic rings are usually immediately glycosylated at the hydroxyl group, and, therefore, isolation of the hydroxylation products is not always possible. However, in particular cases identification of the hydroxylated products is possible. For example, the herbicide bentazon is hydroxylated in the phase of initial transformation into 6-hydroxybentazon or 8-hydroxybentazon. This process is followed by glycosylation, although in plant tissues treated with bentazon besides glycosides the initial herbicide hydroxy derivatives are also found [92, 299, 300]. Plants that are resistant to bentazon transform this herbicide rapidly, whereas the transformation in sensitive plants is carried out slowly [480–482].

The herbicide chlorsulfuron is hydroxylated at the aromatic ring in wheat seedlings and the hydroxyl derivative undergoes direct glycosylation [489]. However, the same herbicide is hydroxylated exclusively at the methyl group of the heterocyclic ring in seedlings of flaxseed (*Linum usitatissimum*) [245].

Herbicides of the sulfonylurea type are usually initially subjected to hydroxylation at the aromatic or heterocyclic ring or at the aliphatic radical, and then the hydroxy derivatives are glycosylated [35]. Thus, the sulfonylurea herbicide primisulfuron is hydroxylated at the pyrimidine ring, but hydroxylation of the benzene ring does not occur in cockspur *(Echinochloa crus-galli)* [353]. On the other hand, microsomes from etiolated maize seedlings catalyze the hydroxylation of this herbicide at both aromatic and heterocyclic rings [172]. In lettuce (*Lactuca sativa*) and vine, not only the aromatic ring and methyl group bind to primisulfuron, but apparently the methyl group of the N-methoxyacetyl radical of metalaxyl is also simultaneously hydroxylated [86].

Finally, the rare hydroxylation at the amide nitrogen must be mentioned. The herbicide phenmedipham undergoes such hydroxylation in leaves of herbicide-resistant and sensitive biotypes of sugar beet, with the degree of transformation in resistant leaves being far higher than in sensitive leaves [100].

Reduction

Organic contaminants containing nitro groups (typically explosives such as TNT, RDX, HMX, etc.) are transformed via reduction. For instance, TNT is initially reduced to ADNTs by most organisms.

The tissue homogenates of rabbit livers, kidneys, or hearts are all capable of reducing TNT. Microorganisms of different taxonomic groups, namely fungi, bacteria and yeast, are also able to reduce TNT [156].

Under aerobic conditions, or conditions with limited oxygen availability, partially reduced nitrotoluenes and secondary condensation products are generated. These processes are completely reduced to TNT under strict anaerobic conditions (Fig. 3.16). TNT and its reduced ADNT congeners are converted during anaerobic treatment to triaminotoluene (TAT), which is chemically unstable [400].

Some strains of *Pseudomonas* and representatives of mycelial fungi are capable of utilizing TNT as a source of nitrogen and carbon, and incorporate atoms of TNT into the skeleton of newly synthesized compounds [156]. This is a good example of how parts of toxic compounds can participate in the creation of compounds vitally important for the organism. *Phanerochaete chrysosporium* and some other basidial fungi completely mineralize TNT. Reduced metabolites of TNT are easily degraded by the oxidative enzymes of basidial fungi. Due to the high intra- and extracellular activities of oxidative enzymes such as lignin peroxidase, Mn-peroxidase, and laccase the strains of basidial fungi have high degradation

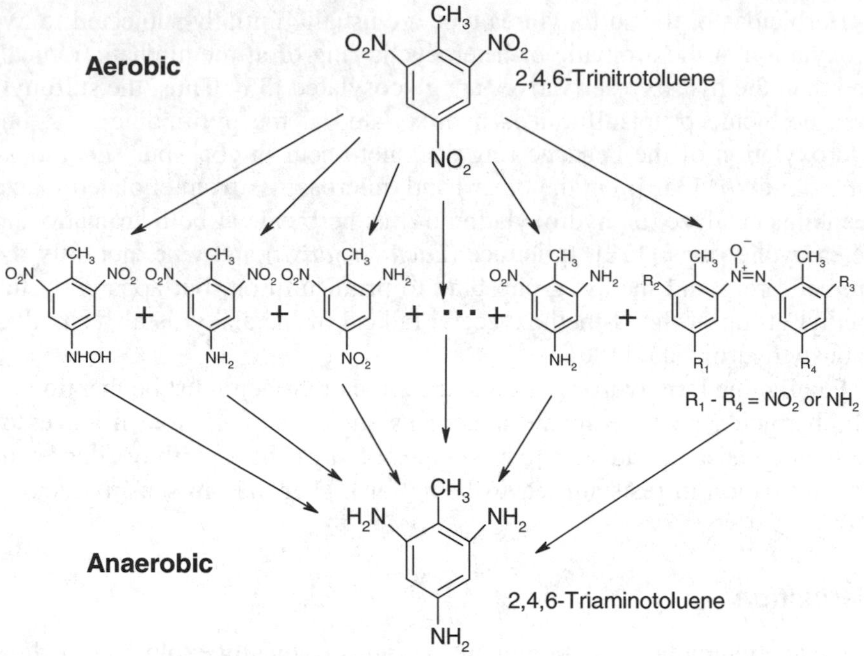

Aerobic

2,4,6-Trinitrotoluene

... (reduction scheme of TNT showing intermediate nitro/amino/hydroxylamino toluene derivatives)

R_1 - R_4 = NO_2 or NH_2

Anaerobic

2,4,6-Triaminotoluene

Fig. 3.16. Reduction of TNT [478]

ability. These fungi are the best microbial detoxifiers of various organic contaminants, including nitroaromatic compounds [156].

Plants are also capable to absorb and assimilate TNT. The aquatic macrophyte parrot feather and the macroalgal stonewort are used for the remediation of TNT-contaminated water [369, 517]. The enzyme nitrore-ductase, which reduces the nitro groups of TNT, is active also in other algae, ferns, monocotyledonous and dicotyledonous plants. Tobacco plants have been genetically engineered to express a bacterial nitroreductase gene, and have acquired the ability to absorb and eliminate TNT from the soil of military proving grounds [213].

Quite a few studies have indicated that TNT disappears from aqueous solutions in the presence of terrestrial and aquatic plants [30–32, 36, 70, 293]. Hexahydro-1,3,5-trinitro-1,3,5-triazine (RDX), another explosive, is absorbed by plants but its degradation is far slower than that of TNT. According to other studies, RDX is stable in solution and accumulates in plant tissues [473]. The ability to take up and metabolize TNT in plants was confirmed by Hughes et al. [243]. They exposed three plant systems, viz. Madagascar periwinkle (*Catharanthus roseus*) hairy root cultures, axenic and native *Myriophyllum* plants, to demonstrate reduction of uniformly labelled [^{14}C] TNT, and to evaluate the fates of the labelled carbon atoms. TNT is completely transformed in all plant systems containing viable plant tissue. The following metabolites have been found: aminonitrotoluenes, some unidentified [^{14}C]-labelled compounds, extractable plant-associated [^{14}C] fractions that could not be identified as reduction products, and bound residues (plant-associated material that could be quantified after combustion of the plant tissue).

TNT can also be reduced to TAT in plants [406], and TAT subsequently undergoes ring cleavage. The enzymes that catalyze reductions of the nitro groups of TNT are nonspecific NAD(P)H-dependent nitroreductases [156]. Note that complete reduction of the nitro groups significantly decreases the mutagenic potential of TNT.

Deep oxidation

The above-presented data summarize the initial transformations of xenobiotics that penetrate into the plant cell. The majority of the low-molecular-mass metabolites formed as a result of such transformations of exogenous molecules are accumulated in vacuoles and the apoplast, often being coupled with endogenous secondary metabolites. Their further transformation is expected to proceed slowly, but this proposition has not been confirmed experimentally.

In experiments on the absorption and transformation of xenobiotics with radioactively labelled carbon atoms, the emission of $^{14}CO_2$ is reported [508], indicating that the initial transformations of xenobiotics is followed by deep oxidation in plant cells.

Plants absorb alkanes and cycloalkanes from the environment and metabolize them. Experiments with [^{14}C]-labelled hydrocarbons proved that sterile seedlings placed in an atmosphere containing low-molecular-mass alkanes (C_1–C_5) or cyclohexane absorb these compounds and further transform them by oxidation to the corresponding carboxyl acids. Alkanes undergo monoterminal oxidation, while cyclohexane is oxidized via ring cleavage. The emission of $^{14}CO_2$ in the dark during this process serves as evidence for the occurrence of mineralization and can be easily measured (depending on the time of exposure, the percentage of mineralization was found to be as high as 30%). Consequently, organic and amino acids are among the end products of this transformation and they can be used for further cell metabolism [371]. For instance, the transformation of methane in tea (*Thea sinensis*) proceeds according to the following scheme (Fig. 3.17):

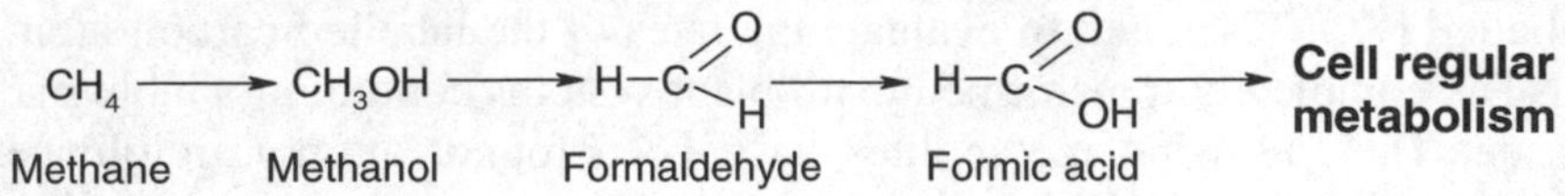

$$CH_4 \longrightarrow CH_3OH \longrightarrow H-C{\overset{O}{\underset{H}{}}} \longrightarrow H-C{\overset{O}{\underset{OH}{}}} \longrightarrow \textbf{Cell regular metabolism}$$

Methane   Methanol   Formaldehyde   Formic acid

Fig. 3.17. Transformation of methane by higher plant cells

Metabolism of ethane, propane and pentane leads to the formation of low-molecular-mass compounds largely composed by organic acids. Labelled fumaric, succinic, malonic, citric and lactic acids have been identified in plant leaves exposed to these low molecular mass alkanes, with most of the radioactivity incorporated into succinic and fumaric acids. Based on the fact that the carbon atoms originating from ethane are incorporated into these acids, it is proposed that ethane in plants is oxidized monoterminally (Fig. 3.18): if ethane was oxidized at both terminal carbon atoms, instead of one, the carbon atoms originating from ethane would be incorporated into glycolic, glyoxalic or oxalic acids. The oxidation of ethane at one terminal carbon atom leads to the formation of acetyl-CoA, which in turn is able to participate in the Krebs cycle [129, 130].

The oxidation of propane at one terminal carbon atom leads to the formation of propionic acid, which successively undergoes β-oxidation resulting in the formation of malonyl-CoA, and decarboxylation resulting in the formation of acetyl-CoA (Fig. 3.19) [508].

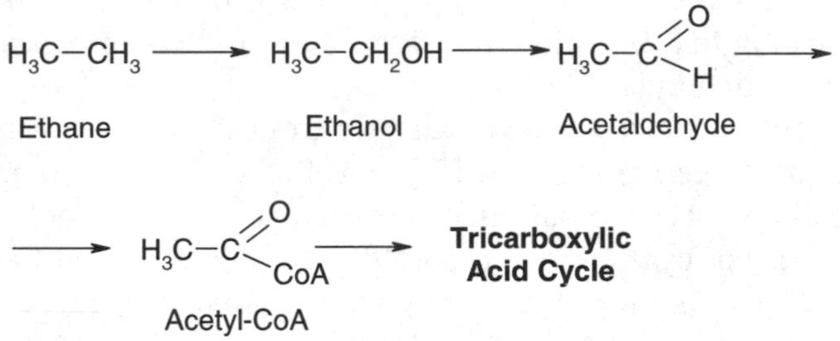

$$H_3C-CH_3 \longrightarrow H_3C-CH_2OH \longrightarrow H_3C-C{\overset{O}{\underset{H}{}}} \longrightarrow$$

Ethane   Ethanol   Acetaldehyde

$$\longrightarrow H_3C-C{\overset{O}{\underset{CoA}{}}} \longrightarrow \textbf{Tricarboxylic Acid Cycle}$$

Acetyl-CoA

Fig. 3.18. Transformation of ethane by higher plant cells

Acetyl-CoA is transferred to carboxyl groups of succinic acid. Based on the identified low molecular mass degradation products, it is suggested that propane is also oxidized monoterminally in plants into compounds that can be incorporated into the Krebs cycle. Pentane may also be oxidized monoterminally, forming valeric acid. Approximately the same organic acids are formed from pentane as from valeric acid [508].

Long-chain alkanes are subjected to transformations similar to those of short-chain alkanes. For instance, after 40 min of incubation of leek leaves with an emulsion of exogenous [^{14}C] octadecane in water, 9.6% of the total label is detected in esters, 6.4% in alcohols, and 4% in organic acids [68]. Following a similar experimental approach, it was firstly demonstrated that plants are also able to metabolize benzene and phenol via aromatic ring cleavage [133, 134, 508]. In this process the carbon atoms are incorporated into carboxyl acids and amino acids. Similar data were obtained for nitrobenzene and aniline [342], toluene [134, 135, 253, 498], α-naphthol [512] and benzidine [136, 137].

Oxidation of benzene and phenol by crude enzyme extracts of plants forms muconic acid after ring cleavage, with catechol as an intermediate, according to the following scheme (Fig. 3.20) [132].

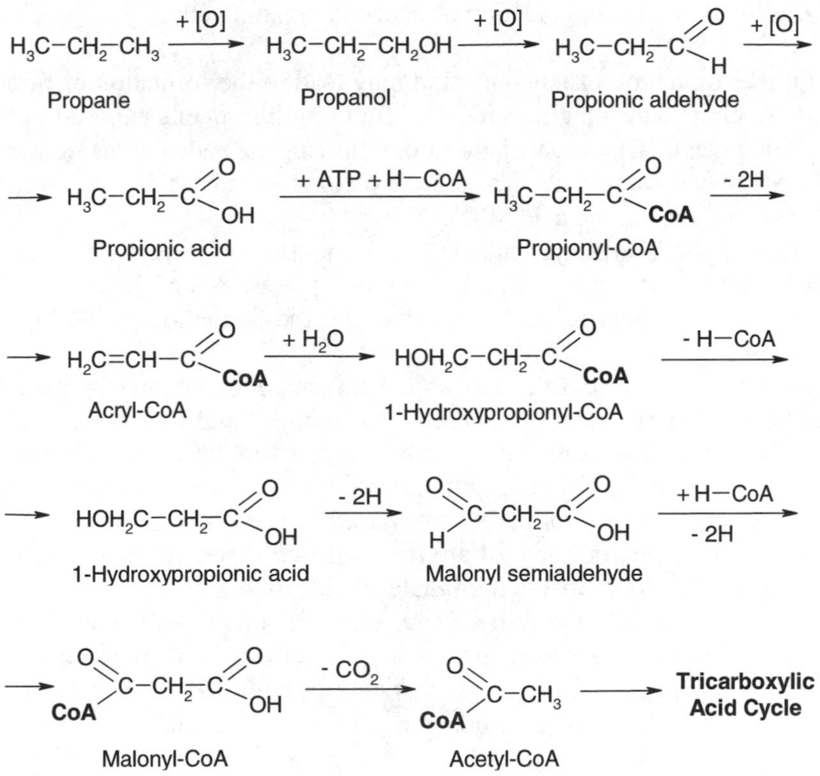

Fig. 3.19. Transformation of propane by higher plant cells

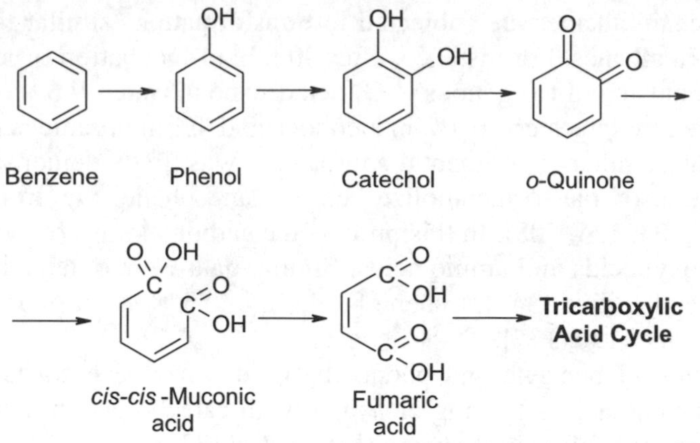

Fig. 3.20. Oxidative degradation of benzene in plant cells

Further oxidation of muconic acid may lead to the formation of fumaric acid. Muconic and fumaric acids are often found in plants exposed to benzene or phenol. Cleavage of the aromatic ring in endogenous substrates proceed in the same way, i.e., 3,4-dihydroxybenzoic acid is transformed into 3-carboxymuconic acid [493].

Phenoxyalkyl-carboxyl acids with 4 and more carbon atoms in their side chain often undergo β-oxidation in plants. For instance, 2,4-dichlorophenoxybutyric acid is oxidized by the formation of 2,4-D [224, 330, 494].

As can be inferred from the above information the role of plants in maintaining an environment tolerable for humans and other fauna is particularly apparent with regard to the burning of fossil fuels for heating, the generation of electricity, and automotive transport (oil products, natural gas, coal, peat, etc.), formerly considered to be fairly harmless, in the course of which carbon oxides are released into the environment in sufficient quantities to swiftly exterminate all life if plants were not actively removing them from the atmosphere. Green plants together with certain microorganisms form the main naturally beneficial ecological power, directed towards maintaining the atmosphere in a tolerable condition by absorbing and metabolizing carbon monoxide and dioxide and other contaminants. Yet, the current global warming has made it apparent that the biological self-cleaning potential of our planet is now exceeded by the rate of accumulation of pollutants, as evidenced not only by adverse changes in climatic parameters in the most damaged regions, but also in the appearance of defective forms of plants, animals and microorganisms.

3.3.3 Enzymes transforming organic contaminants

Organic contaminant degradation processes are closely related to cellular metabolism in many ways. In prolonged detoxification processes many enzymes are involved. These enzymes may be divided into two groups: enzymes directly and indirectly participating in detoxification processes.

Reactions occurring during all three detoxification processes (functionalization, conjugation and compartmentation) are of enzymatic nature. In the absence of xenobiotics these enzymes catalyze other reactions typical for regular plant cell metabolism. The following enzymes directly participate in the initial modification of organic contaminants:

- Oxidases, catalyzing hydroxylation, demethylation and other oxidative reactions (cytochrome P450-containing monooxygenases, peroxidases, phenoloxidases, ascorbatoxidase, catalase, etc.).
- Reductases, catalyzing the reduction of nitro groups (nitroreductases).
- Dehalogenases, splitting atoms of halogens from halogenated and poly-halogenated xenobiotics.
- Esterases, hydrolyzing ester bonds in pesticides and other organic contaminants.

As has already been mentioned, conjugation is catalyzed by transferases (GST, glucuronosyl-O-transferase, etc). Compartmentation of conjugates takes place with the participation of ATP-binding cassette (ABC) transporters [141]. Depending on the structure of the xenobiotic, other enzymes may also participate at different stages of the intracellular oxidation of the contaminant. For instance, conjugates containing glutathione and deposited in vacuoles are transformed by peptidases into conjugates containing only cystein [87].

During deep oxidation in plants, enzymes involved in the processes that provide the plant cell with energy, and which are important for defense reactions by the provision of secondary metabolites, etc., indirectly participate in the detoxification processes.

The main enzymatic reactions that provide functionalization of the organic contaminants in the plant cell are presented below in Figs. 3.21–3.37.

$$H_3C-CH_3 \longrightarrow H_3C-CH_2-OH$$

Ethane   Ethanol

Fig. 3.21. Aliphatic hydroxylation of ethane

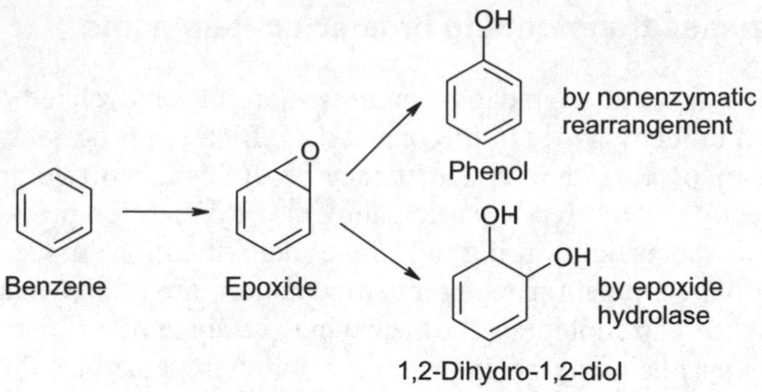

Fig. 3.22. Aromatic hydroxylation (via epoxide)

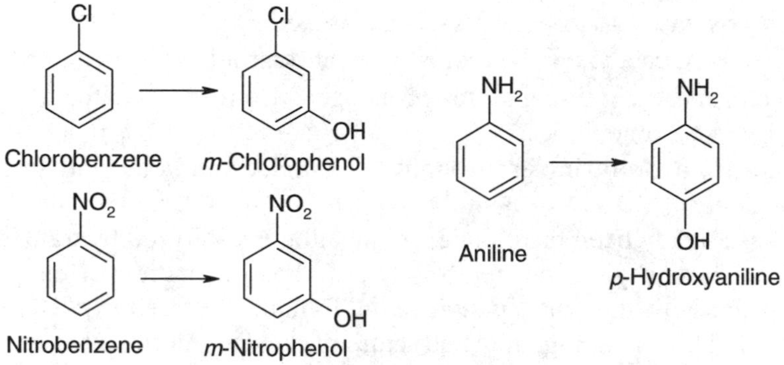

Fig. 3.23. Aromatic hydroxylation (via O-insertion)

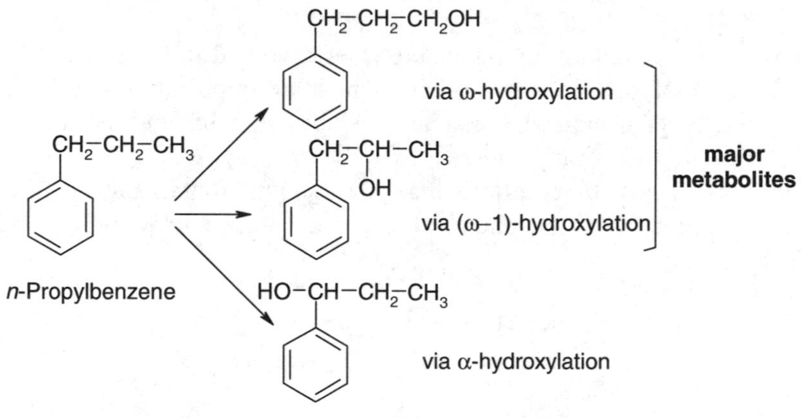

Fig. 3.24. Aliphatic hydroxylation of n-propylbenzene

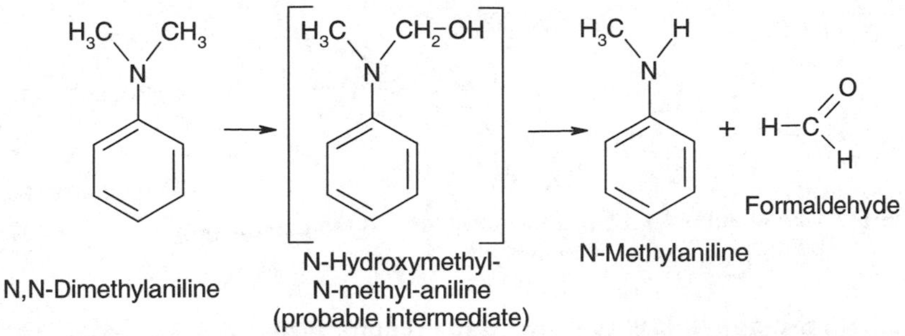

Fig. 3.25. N-dealkylation with oxidation of carbon atom of the alkyl group

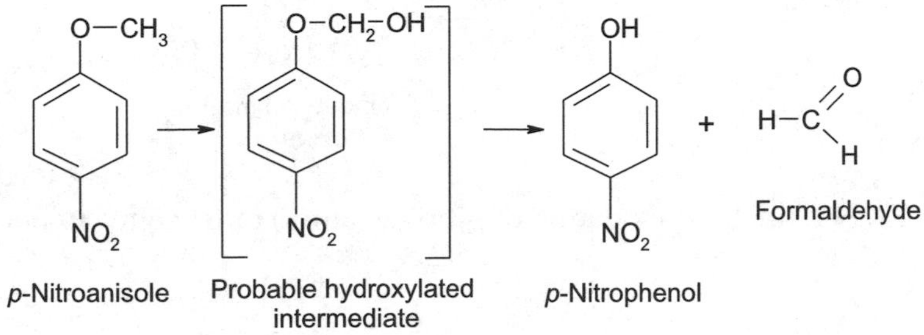

Fig. 3.26. O-dealkylation of aromatic ethers

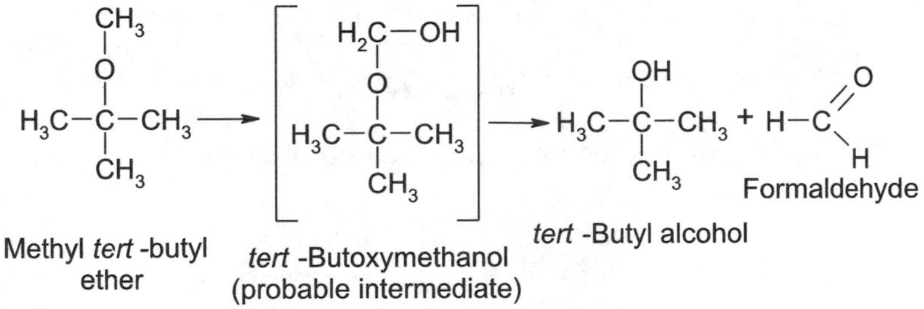

Fig. 3.27. O-dealkylation of aliphatic ethers

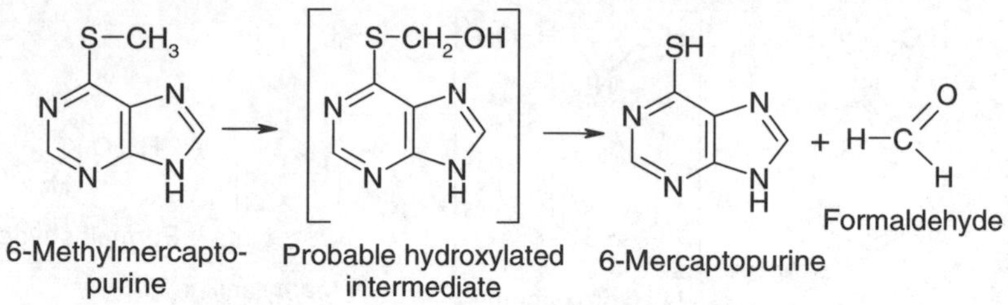

Fig. 3.28. S-dealkylation of 6-methylmercaptopurine

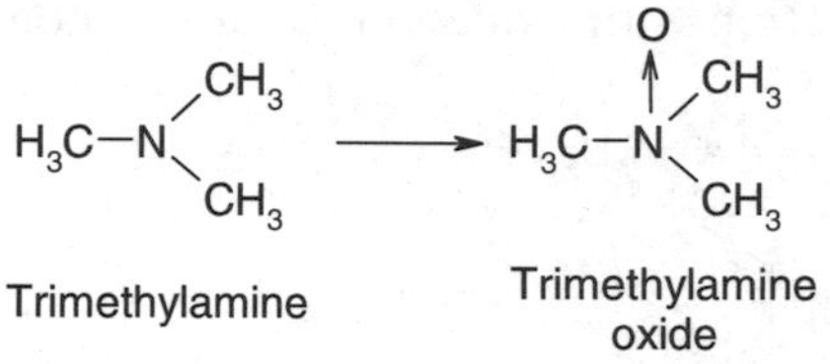

Fig. 3.29. N-oxidation (oxidation of nitrogen atom) of trimethylamine

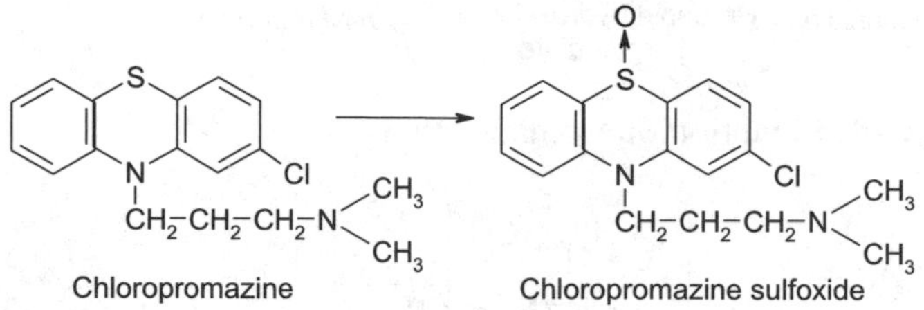

Fig. 3.30. S-oxidation (oxidation of sulfur atom) of chloropromazine

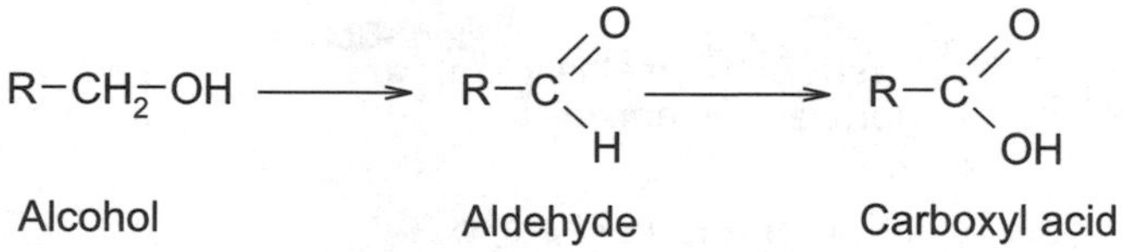

Fig. 3.31. Alcohol and aldehyde oxidation

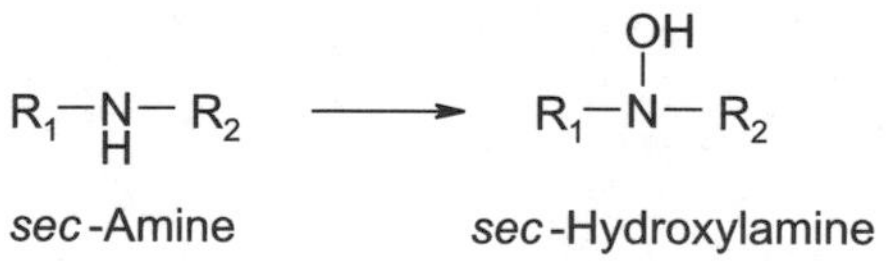

Fig. 3.32. N-hydroxylation of *sec*-amines

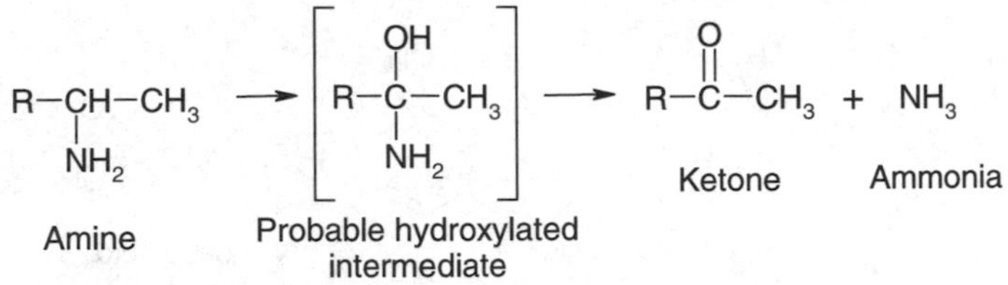

Fig. 3.33. Oxidative deamination of *sec*-amines

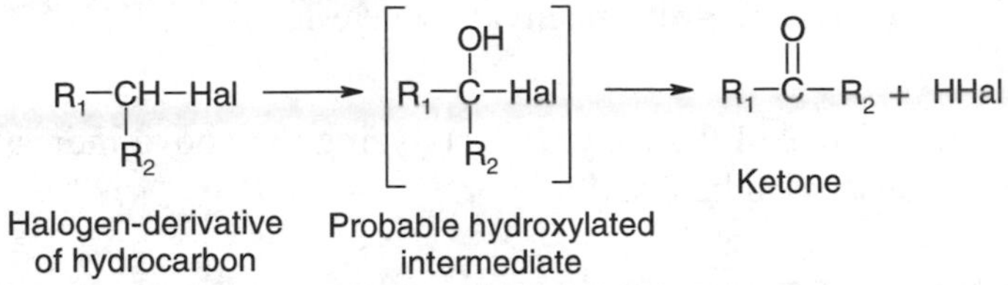

Fig. 3.34. Oxidative dehalogenation of *sec*-halogenides

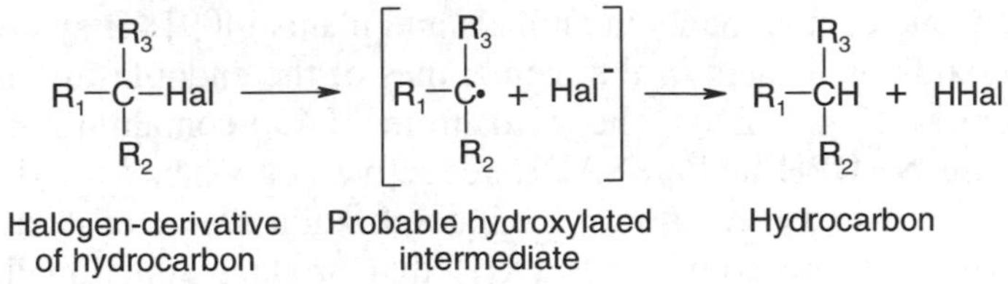

Fig. 3.35. Reductive dehalogenation of *tert*-halogenides

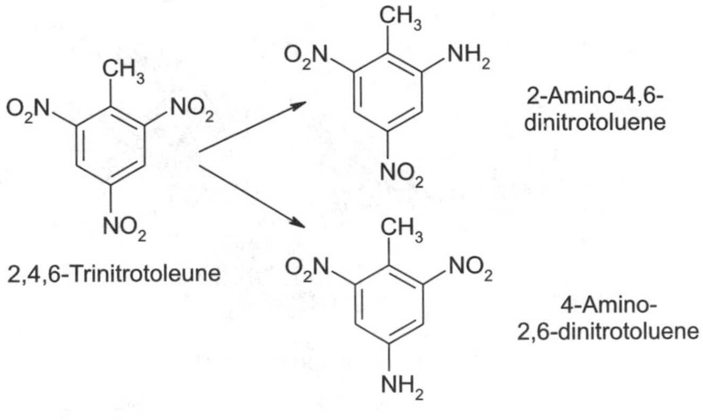

Fig. 3.36. Reduction of TNT nitro groups

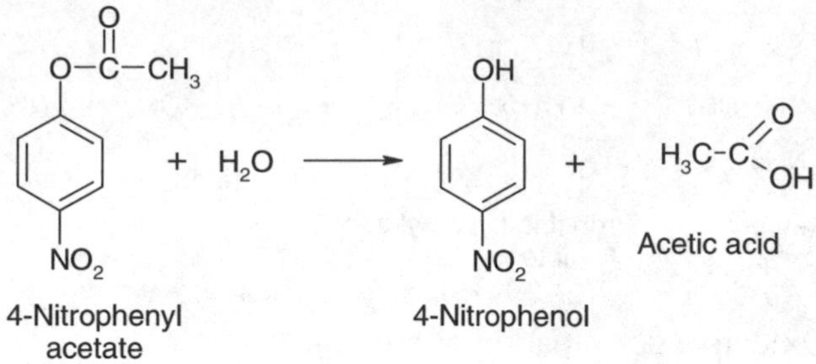

Fig. 3.37. Hydrolysis of 4-nitrophenyl acetate ester

The characteristics of the enzymes catalyzing the above indicated transformation are discussed below.

Cytochrome P450-containing monooxygenases

Cytochrome P450-containing monooxygenases (EC 1.14.14.1) belong to one of the major groups of enzymes that are responsible for the detoxification of organic contaminants in animals and plants [409]. They are mixed-function oxidases located in the membranes of the endoplasmic reticulum (microsomes) (Fig. 3.38). The cytochrome P450-containing monooxygenases use NADPH and/or NADH reductive equivalents for the activation of molecular oxygen and for the incorporation of one of its atoms into hydrophobic organic compounds (XH) that produce functionalized products (XOH) [443]. In this case the second atom of oxygen is used for the formation of a water molecule.

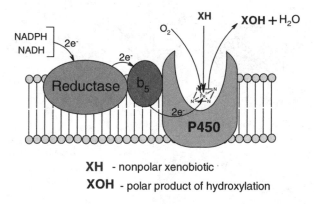

XH - nonpolar xenobiotic

XOH - polar product of hydroxylation

Fig. 3.38. Microsomal monooxygenase system

The microsomal cytochrome P450-containing monooxygenase system is an electron transfer chain. All constituents of this multicomponent system are located in the membranes of the endoplasmic reticulum. The monooxygenase system contains the following components: (1) the initiator of electron transfer, a NADPH-cytochrome P450 reductase (EC 1.6.2.4); (2) the intermediate carrier—cytochrome b_5; and (3) the terminal acceptor of electrons—cytochrome P450. When NADPH is used as the only source of reducing equivalents in this system, the existence of an additional carrier, a NADH-dependent flavoprotein, is required. NADH may also be oxidized by the NADPH-dependent redox system. In the latter case, cytochrome b_5 is not needed as the intermediate carrier [215].

The cytochrome P450-dependent hydroxylation process consists of the following steps (Fig. 3.39) [232]:
1. The process begins when the xenobiotic (XH) binds to the active site of the oxidized cytochrome P450 and an enzyme-substrate complex, called ferricytochrome P450, is formed. If the reaction progresses further, the water molecule that forms a ligand of the iron atom of the heme in the

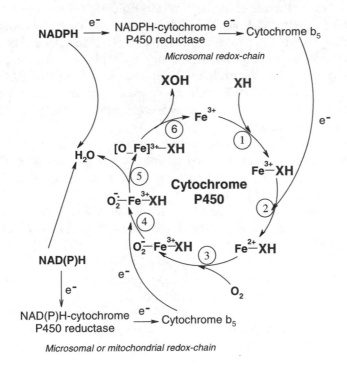

Fig. 3.39. Organic compound hydroxylation by cytochrome P450-containing monooxygenase

unbound cytochrome P450 is displaced. This is accompanied by a change in the spin of the Fe^{3+} from a low spin = 1/2 state, in which the $3d^5$ electrons are maximally paired, to a high spin = 5/7 state, when the electrons are maximally unpaired. This, in turn, causes a change in the redox potential of the iron from approximately −300 mV to −170 mV. This change is sufficient to render the reduction of iron by the redox-partner of the cytochrome, usually NADPH or NADH, thermodynamically favorable.

2. The one-electron transfer from NADPH yields the reduction of the enzyme-substrate complex, leading to the formation of ferrocytochrome P450. This reaction is catalyzed by the NADPH-cytochrome P450 reductase.

3. The reduced cytochrome P450-substrate complex reacts with molecular oxygen and forms the ternary complex oxycytochrome P450. This state is not stable and is easily autooxidized, releasing O_2. However, if the transfer of a second electron occurs (see next step), the catalytic reaction continues.

4. The ternary complex undergoes a second one-electron reduction. This step is rate limiting for the overall process of cytochrome P450-dependent hydroxylation. The reduction of the oxycytochrome P450-substrate complex to peroxycytochrome P450, takes place. The donor of the second one-electron transfer may vary with the substrate, or availability of the reduced pyridine nucleotides, or with both of these factors. NADPH-cytochrome P450 reductase appears predominant, but the electron may be provided also by microsomal or mitochondrial NADH cytochrome b_5 reductase.

5. Peroxycytochrome P450 decomposes releasing water. Then the O_2^{2-} reacts with protons from the surrounding solvent to form H_2O that is released, and, thus, in the active site of the enzyme one activated oxygen atom is left.

6. The final step is the release of the hydroxylated substrate and the oxidized cytochrome P450. At this stage the oxidized hemoprotein can be recycled by becoming bound to another substrate molecule.

This entire reaction cycle usually takes 1 to 10 seconds.

Cytochrome P450-containing monooxygenase systems primarily fall into two major classes: bacterial/mitochondrial (type I), and microsomal (type II). Alternatively, cytochrome P450-containing systems can be classified according to the number of protein components. Mitochondrial and most bacterial cytochrome P450-containing systems have three components: an FAD-containing flavoprotein (NADPH or NADH-dependent reductase), an iron-sulfur protein, and cytochrome P450. The eukaryotic microsomal monooxygenase system contains two components: NADPH-cytochrome P450 reductase (a flavoprotein containing both FAD and FMN) and cytochrome P450. A soluble monooxygenase P450 BM-3 from *Bacillus megaterium* exists as a single polypeptide chain with two functional parts (the heme and flavin domains), and represents a unique bacterial one-component system. The sequence and functional comparison show that these domains are more similar to cytochrome P450 and the flavoprotein of the microsomal two-component cytochrome P450 monooxygenase system than to the relevant proteins of the three-component system [232].

Specialized features distinguish the organization and functioning of the cytochrome P450-containing monooxygenase systems in prokaryotic and eukaryotic organisms. Prokaryotes contain soluble forms of this enzymatic system. In eukaryotes, the structure of the hemoproteins is established by their incorporation into the endoplasmic membrane [14]. The classic example is liver cytochrome P450, which is readily incorporated into the membrane structure. The individual components of the monooxygenase system are located along the entire membrane. In this configuration they are in close contact with the lipid matrix, which at the same time has a barrier function. Oxidative hydroxylation in microsomes therefore takes place preceded by the penetration of the xenobiotic through the membrane lipid layer. Formation of a catalytically active complex between cytochrome P450 and the xenobiotic determines its movement from the aqueous to the phospholipid phase.

Cytochromes P450 are universally distributed and are detected in animals, plants and microorganisms. Cytochromes P450 are encoded by a highly divergent gene superfamily, and exhibit great diversity in reactive site and amino acid composition [443]. This superfamily contains a spectrum of CYP gene families, which differ substantially in their primary sequence, substrate specificity, genomic organization and inducibility. Over 120 cDNA and genomic DNA sequences for P450s of different plants have been identified [443]: e.g., wheat, avocado *(Persea americana),* aubergine *(Solanum melongena* cv. Sinsadoharanasu), catmint *(Nepeta racemosa),* Madagascar periwinkle, peppermint *(Mentha piperita),* pennycress *(Thlaspi arvense),* thale cress, maize, Jerusalem artichoke *(Helianthus tuberosus),* mung bean, alfalfa, sunflower, pea, flaxseed, guayule *(Parthenium argentatum),* petunia *(Petunia hybrida),* moth orchid *(Phalaenopsis* sp. hybrid SM9108), sorghum, barberry *(Berberis stolonifera),* field mustard *(Brassica campestris),* pigeon pea, tobacco, soybean, etc.

Plants, in contrast to animals, possess a cytochrome P450-containing monooxygenase system built into the membrane. However, plant cells also contain soluble forms of the same enzyme, enabling them to significantly enhance their detoxifying capacity [195].

Cytochromes P450 catalyze extremely diverse and often complex regio-specific and/or stereospecific reactions in the biosynthesis or catabolism of regular and secondary metabolites of plant [345]. There are over 20 physiologically important processes and reactions in which cytochrome P450 plays a key role [140, 443]. Those of vital importance for the plant cell include: biosynthesis of lignin monomers [527], anthocyanins [235], furanocoumarins [28], gibberellins [255], isoflavonoid phytoalexins [273], alkaloids [286], hydroxylation of fatty acids [421], hydroxylation of limo-nene and geraniol [210], etc. At the same time some plant cytochrome P450-containing monooxygenases can play an important role in the hy-droxylation of exogenous toxic compounds (environmental contaminants and other xenobiotics) after they penetrate into the plant cell [428]. Plant cytochromes P450 participate in the reactions of C- and N-hydroxylation of aliphatic and aromatic compounds, N-, O-, and S-dealkylation, sulfo-oxidation, deamination, N-oxidation, oxidative and reductive dehalogena-tion, etc. [443]. The resistance against many herbicides in plants is medi-ated by the rapid transformation of the herbicide into a hydroxylated, inac-tive product that is subsequently conjugated to carbohydrate moieties in the plant cell wall. For example, N-demethylation and ring-methyl hy-droxylation of the phenylurea herbicide chlorotoluron in wheat and maize are cytochrome P450-dependent processes [172, 348]. After hydroxylation both products undergo conjugation with glucose via the newly formed hy-droxyl group [348].

For some phenylurea herbicides cytochrome P450-mediated N-demethylation in the Jerusalem artichoke is sufficient to cause partial or complete loss of phytotoxicity (Fig. 3.40) [115].

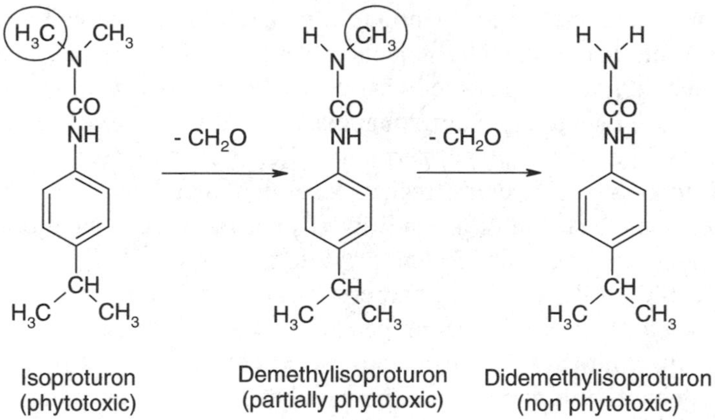

Isoproturon
(phytotoxic)

Demethylisoproturon
(partially phytotoxic)

Didemethylisoproturon
(non phytotoxic)

Fig. 3.40. N-Demethylation of isoproturon by cytochrome P450-dependent monooxygenase

Sulfonylurea herbicides (primisulfuron, chlorsulfuron and triasulfuron) in wheat and maize undergo hydroxylation in the aromatic ring under the action of cytochrome P450-dependent monooxygenase (Fig. 3.41) [443].

Analogously, cytochrome P450-catalyzed hydroxylations are typical for other aromatic ring-containing herbicides. For example, diclofop in wheat undergoes such transformations (Fig. 3.42) [331], and bentazon in maize (Fig. 3.43) [332]. Similarly to chlorotoluron metabolites, the products of diclofop and bentazon are conjugated to O-glucosides after the hydroxylation.

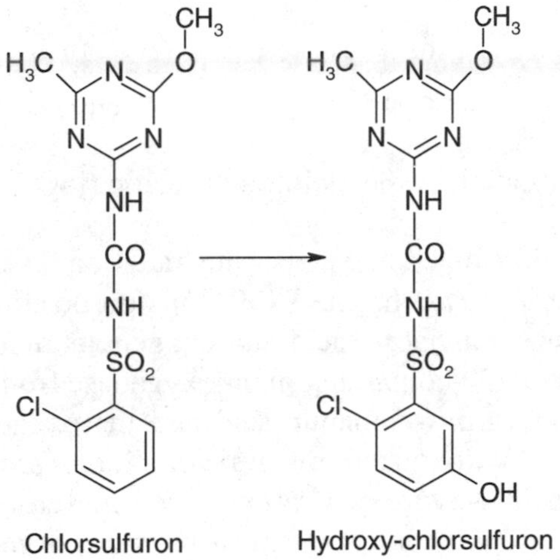

Chlorsulfuron   Hydroxy-chlorsulfuron

Fig. 3.41. Hydroxylation of aromatic ring of chlorsulfuron

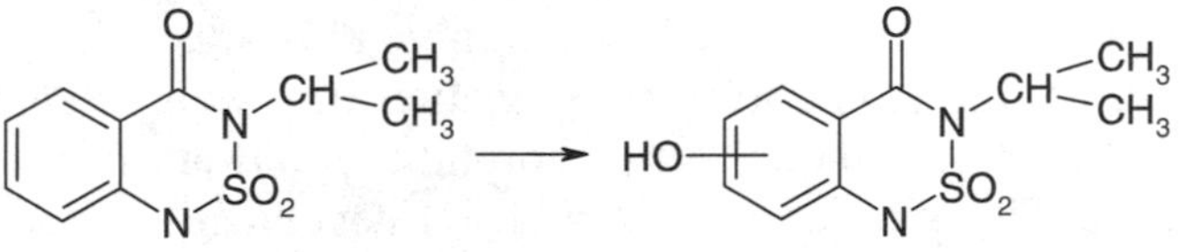

Bentazon   Hydroxy-bentazon

Fig. 3.42. Hydroxylation of the bentazon aromatic ring

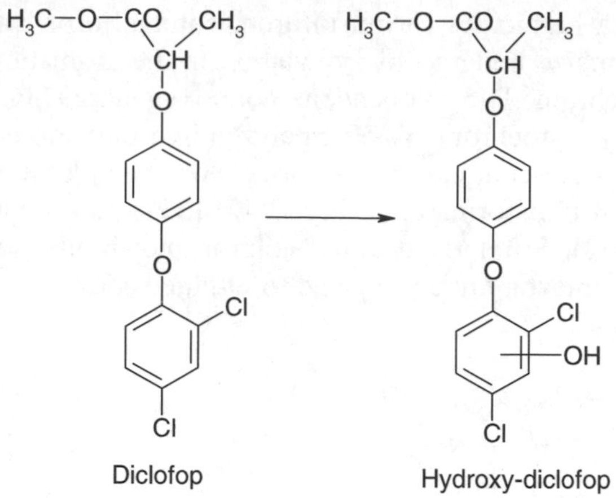

Diclofop   Hydroxy-diclofop

Fig. 3.43. Hydroxylation of the diclofop aromatic ring

The hydroxylation of endogenous substrates and xenobiotics may be catalyzed by the same cytochrome P450. This proposition is supported by the fact that endogenous lauric acid and exogenous diclofop are oxidized by the cytochrome P450-containing monooxygenase from wheat [549], and that the endogenous *trans*-cinnamic acid and the exogenous p-chloro-N-methylaniline are hydroxylated by a recombinant artichoke CYP73A1 (*trans*-cinnamic acid-4-hydroxylase) expressed in yeast [382]. During simultaneous incubation of a microsomal suspension (from etiolated soybean seedlings) with [1-^{14}C] *trans*-cinnamic acid (as endogenous substrate) and N,N-dimethylaniline (as model xenobiotic), the hydroxylation of the endogenous substrate was inhibited up to 70–80% [194]. On the other hand, the demethylation of N,N-dimethylaniline in the presence of *trans*-cinnamic acid was inhibited by only 25–30%. Besides N,N-dimethylaniline, the enzymatic transformation of cinnamic acid was also inhibited by other xenobiotics (ethylmorphine, p-nitroanisole, aniline and aminopyrine). The kinetics of the NADPH-dependent oxidation of cinnamic acid and xenobiotics revealed the competitive character of the inhibition of the cinnamic acid–hydroxylase activity by xenobiotics [266]. These results indicate that there is a switching of cytochrome P450 to hydroxylation of xenobiotics enabled by and due to a decrease in its physiological function (hydroxylation of *trans*-cinnamic acid). The switch of an enzyme from biosynthesis to detoxification is determined by the polarity of the xenobiotic: the more hydrophobic the xenobiotic, the higher its affinity for cytochrome P450, the more universal the switch, and the faster the process of xenobiotic oxidation proceeds. Thus, it seems that after penetration of hydrophobic xenobiotics into the plant cell, switching of cytochrome P450 from an "endogenous" to an "exogenous" function regimen may take place. In essence, such switching is set into motion by the superior affinity of the xenobiotic for the enzyme compared to that of its natural substrates [196].

In plants growing in a medium containing a xenobiotic, the concentration of cytochrome P450 increases. Nearly all xenobiotics examined have an inductive nature. Inductive abilities of various xenobiotics such as phenobarbital, clofibrate, aminopyrine, and the herbicides 2,4-D, propanil, chloroacetamide, thiocarbamate, chlorotoluron, bentazon and others have been described [420]. A cytochrome P450 (CYP76B1) isolated from Jerusalem artichoke is more actively induced by xenobiotics than other cytochrome P450-containing monooxygenases. This CYP76B1 (**CYP** designates **cytochrome P**450, **76** designates the gene family, **B** designates the gene subfamily and **1** designates a particular gene) [75] metabolizes with high efficiency a wide range of xenobiotics, including alkoxycoumarins, alkoxyresorufins, and several herbicides of the phenylurea class [409]. CYP76B1 catalyzes also the removal of both N-alkyl groups of phenylureas with turnover rates comparable to physiological substrates and produces non-phytotoxic compounds. This cytochrome P450-increased herbicide metabolism and tolerance can be achieved by ectopic constitutive expression of CYP76B1 in tobacco and *Arabidopsis* [115]. Transformation with CYP76B1 resulted in a 20-fold increase in tolerance to the herbicide linuron and a 10-fold increase in tolerance to the herbicides isoproturon or chlorotoluron in tobacco and *Arabidopsis*. Other than increased herbicide tolerance, the expression of CYP76B1 results in no other visible phenotypic change in the transgenic plants. CYP76B1 can function as a marker for plants that can be selected for the phytoremediation of contaminated sites.

The inductive effect of each particular xenobiotic depends on its chemical nature and/or the inductive potential of its intermediates. Some of these intermediates appear to be highly reactive and most of them cause inactivation of cytochrome P450 and its further conversion into cytochrome P420. Good examples of these intermediates are metabolites of 3,4-benzopyrene. Incubation of soybean and ryegrass with 3,4-benzopyrene causes the formation of epoxides, dioles and quinones [427, 503]. The aggressiveness of these substances is expressed by the formation of active oxygen radicals that cause the irreversible conversion of cytochrome P450 to P420 [266]. For instance, enhancement of the peroxidation of fatty acids also leads to the generation of oxygen radicals. According to hitherto unpublished data of the present authors, in maize seedling microsomes, superoxide anion radicals are generated during cytochrome P450-mediated 3,4-benzopyrene oxidation.

The expression of genetically engineered cytochrome P450 would be required for the low-cost production of several natural products, such as antineoplastic drugs (taxol or indole alkaloids), nutraceuticals (phytoestrogens) and antioxidants in plants [345]. These compounds may have important functions in plant defense. Engineered cytochrome P450s could improve plant defense against insects and pathogens. These P450s may be tools to modify herbicide tolerance, and are selectable markers for phytoremediation.

Peroxidases

Peroxidases (EC 1.11.1.7) are ubiquitous enzymes found in virtually all green plants, the majority of fungi and aerobic bacteria. The isoenzyme heterogeneity of peroxidases appears to be the result of *de novo* synthesis, as well as an array of physiological and ecological determinants including hormones, light, infection, etc. [465]. Peroxidases have phylogenetically correlated similarities based on the chemical nature and redox potentials of the substrates that they oxidize. Peroxidases often increase in response to stress, and one of the principal roles of peroxidases appears to be the protection of cells from hydrogen peroxide. The great catalytic versatility of peroxidase is its predominant characteristic, and, therefore, no single role exists for this multifunctional enzyme.

Peroxidase activity is defined by the following reaction:

$$RH_2 + H_2O_2 \rightarrow 2H_2O + R$$

Peroxidase is composed of a single peptide chain, and contains one heme (protoporphyrin IX). The plant peroxidases (as distinct from those of animals) contain approximately 25% of carbohydrates that protect the enzyme from the action of proteolytic enzymes, and stabilize the protein conformation [242].

The peroxidases are known to catalyze a number of free radical reactions [478]. The resting enzyme (ferric-heme protein), is initially oxidized by two electrons in a reaction with hydrogen peroxide (Fig. 3.44). An early step in the catalytic cycle following the binding of hydrogen peroxide to the heme in the Fe(III) state is the heterolytic cleavage of the O–O bond in hydrogen peroxide. Two key catalytic residues in the distal heme pocket, an arginine (Arg_{38}) and a histidine (His_{42}), are involved in the activation of peroxide and the formation of compound I by catalyzing the transfer of a proton from the α- to the β-oxygen atom of heme-bound H_2O_2 and polarizing the O–O bond. His_{42} at a neutral pH acts initially as a proton acceptor (base catalyst) and then as a donor (acid catalyst). Arg_{38} is influential in lowering the pK_a of

His$_{42}$ and additionally in aligning H_2O_2 in the active site, but it does not play a direct role in the proton transfer [413]. The resulting intermediate, compound I, is an oxyferryl Fe(IV) heme protein with a cation radical stabilized in the heme porphyrin ring. Compound I is then converted back to the resting enzyme via two successive single-electron transfers from reducing substrate molecules. As a result of the first reduction a second enzyme intermediate (compound II), which retains the heme in the ferryl state (ferryl enzyme), is formed from the porphyrin radical cation. Compound II can be reduced again to regenerate ferric enzyme or react with hydrogen peroxide to form a catalytically inactive species, compound III (ferric-superoxide protein). Alternatively, the compound that is directly oxidized by the enzyme oxidizes other organic compounds, including xenobiotics. Analogously, peroxidases of microorganisms participate in oxidative reactions (lignin peroxidase and Mn-dependent peroxidase). In addition to the oxidation of different compounds, lignin peroxidase also catalyzes reductive reactions in the presence of electron donors such as EDTA and oxalate [478].

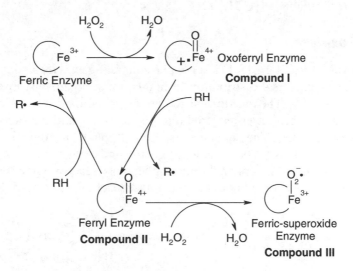

Fig. 3.44. The action pattern of peroxidase

The electron donors, such as veratryl alcohol (a free radical mediator) are oxidized by lignin peroxidase and generate cation radicals. As a result, the anion radical can catalyze the reduction of electron acceptors (cytochrome c, nitroblue tetrazolium, O_2). Similar reactions have been observed with Mn-dependent peroxidase in the presence of quinones. These reductive mechanisms may be involved in the metabolism of TNT in case of *Phanerochaete chrysosporium* action, but it is shown that these peroxidases do not participate in the initial reducing steps of this explosive [478].

According to the current view nearly all kinds of organic contaminants in plants are oxidized by peroxidases [483]. This notion is based on such facts as the ubiquitous occurrence of the enzyme in plants (the isozymes of peroxidase in green plants occur in the cell walls, plasmalemma, tonoplasts, and intracellular membrane systems of the endoplasmic reticulum, plastids and cytoplasm), the high affinity of peroxidases from different plants to organic xenobiotics of different chemical structures, and their low substrate specificity. These facts indicate the universal character of the enzyme's action and the active participation of peroxidase isoenzymes in a wide variety of detoxification processes. The results of many investigations also indicate the participation of plant peroxidases in hydroxylation reactions of xenobiotics. For example, peroxidases from different plants are capable of oxidizing N,N-dimethylaniline [459], 3,4-benzopyrene, 4-nitro-*o*-phenylendiamine [534], 4-chloroaniline [294], phenol, aminofluorene, acetaminophen, diethylstilbestrol, butylated hydroxytoluene, hydroxyanisoles, benzidine, etc. [428]. According to some data, horseradish (*Armoracia rusticana*) peroxidase can oxidize tritium-labelled [C^{3}H$_3$] TNT [2].

Phenoloxidases

Phenoloxidase (EC 1.14.18.1) is a copper-containing enzyme and is characterized by its universal distribution in plants, microorganisms, insects and animals [329, 486]. The reaction carried out by phenoloxidase is of great importance in such processes as vertebrate pigmentation and the browning of fruits and vegetables. The enzyme exists in multiple forms in active and latent forms, and catalyzes both monooxygenase and oxygenase reactions: the *o*-hydroxylation of monophenols (monophenolase reaction) and the oxidation of *o*-diphenols to *o*-quinones (diphenolase reaction) (Fig. 3.45) [425].

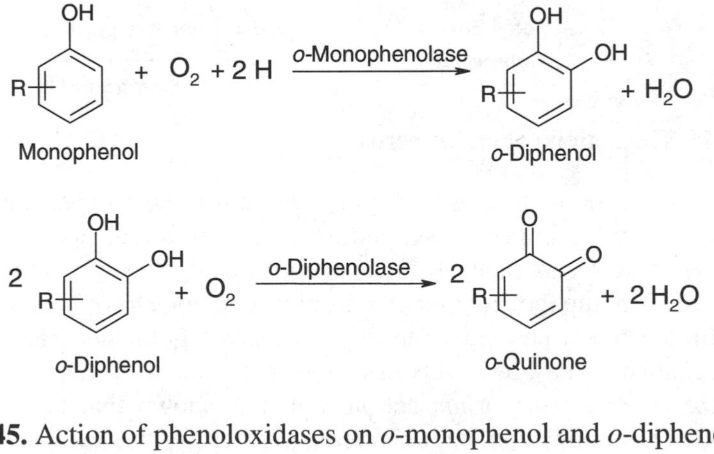

Fig. 3.45. Action of phenoloxidases on *o*-monophenol and *o*-diphenol

The catalytic cycles for the monophenolase and diphenolase activities are coupled not only to each other, but also to non-enzymatic reactions involving *o*-quinone products [411].

The tentative mechanism of the phenoloxidase action is based on results determined for tyrosinase from the basidial fungus *Agaricus bisporus*. The great majority of strains representing this genus are active producers of phenoloxidases [412]. The enzyme has three forms, met-, deoxy- and oxy-, depending on the state of two copper ions of the binuclear site, where they are surrounded by six nitrogen atoms of histidine residues (Fig. 3.46). The met- and oxy- form copper ions are bivalent, and the deoxy- form copper ion is univalent. Besides, substrates (e.g., *o*-diphenol) bind to the met- and oxy- forms, but not to the deoxy- form. Oxygen can only bind to the free deoxy- form.

In diphenolase activity, the *o*-diphenol binds to the axial position of one (left) of the copper (II) ions of the met site (E_{met} in the Fig. 3.46). Coordination of the *o*-diphenol is accompanied by the transfer of a proton to a protein residue (represented in the scheme by bidentate coordination of the *o*-diphenol) and is further accompanied by a second proton transfer,

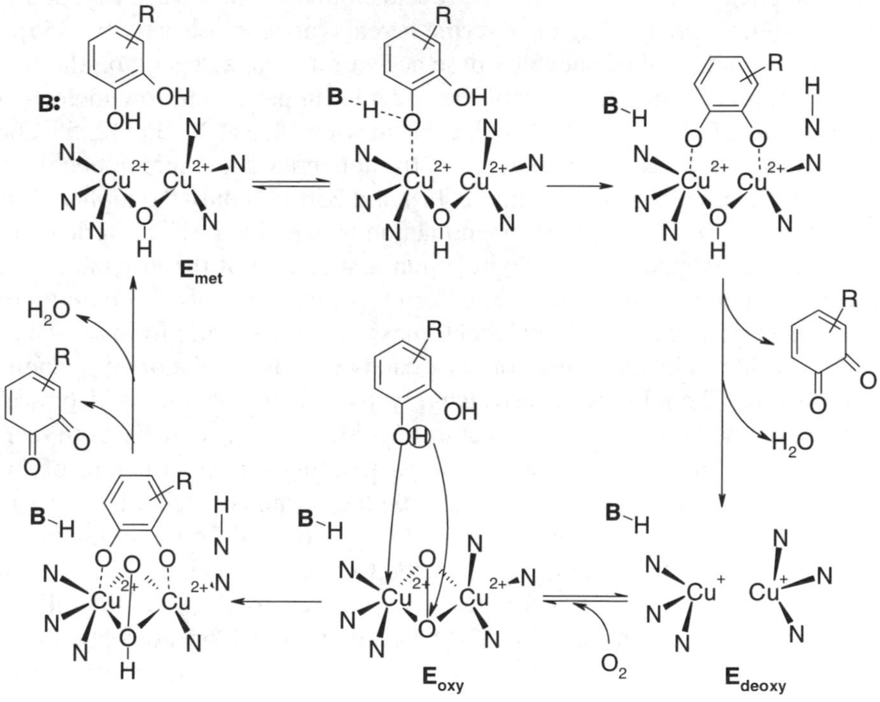

Fig. 3.46. The mechanism of phenoloxidase action

probably enabled by displacement of an axial histidine coordinated to the second (right) copper (II) ion. Electron transfer from the o-diphenol substrates results in formation of the o-quinone and the deoxy- form (E_{deoxy}) of the binuclear copper site, where the metal ions are in the monovalent state.

In the neutral zone, which is provided by the protonated acid–base catalyst, rapid binding of oxygen molecule to the deoxy- form of phenoloxidase could be favored. In the oxy- form the atoms of oxygen are proposed to bind in the peroxide form. Further binding of the o-diphenol to one copper (II) ion of the oxy- form takes place. After binding, the phenoloxidase oxidizes the substrate to o-quinone. That stage is the rate-limiting step of the complete catalytic cycle. It is accompanied by the transfer of a proton to the bound peroxide. Bidentate coordination of the o-diphenol substrate is accompanied by a second proton transfer. The transfer of an electron from the o-diphenol to the peroxide induces the cleavage of the O–O bond to form a water molecule and the o-quinone while regenerating the initial met- form of phenoloxidase. In this last step the protein residue B acts as an acid, providing a proton for the release of water [412].

As mentioned above, the oxy-form of phenoloxidase binds both o-diphenol and monophenol. As shown in the scheme (Fig. 3.47), in the case of monophenol binding the enzyme reveals monophenolase activity, and in case of o-diphenol binding the enzyme reveals diphenolase activity [356].

The type and rate of phenoloxidase activity in plants depend on the molecular mass of multiple forms of the enzyme. In particular, low molecular mass forms of phenoloxidase (molecular masses 14, 21, 28, 35, 42, 55, and 70 kD) expose the abilities to act as both diphenolase and monophenolase; with an increase in molecular mass (118 and 250 kD) only the diphenolase activity is retained [389]. This phenomenon is explained by the steric overlap of the active site of the enzyme upon association of the low molecular forms to create the high molecular forms, which prevents the binding of monophenols with the high molecular mass, i.e., oligomeric forms.

In addition to its main function, i.e., catalyzing the oxidation of phenolic compounds, phenoloxidase also actively participates in the oxidation of xenobiotics with an aromatic structure. In this process, actually both, peroxidase and phenoloxidase are active, depending on the structure of the substrate. Phenoloxidase from spinach oxidizes aromatic xenobiotics (benzene, toluene), and is active in their hydroxylation and further oxidation to quinone [515]. If the xenobiotic subjected to oxidation is not a substrate of the phenoloxidase, such a xenobiotic undergoes co-oxidation in the following manner: the enzyme oxidizes the corresponding endogenous phenol by forming quinones or semi-quinones or both, i.e., compounds with a high redox potential. These compounds activate molecular oxygen and form oxygen radicals, such as a superoxide anion radical ($O_2^{\cdot-}$) and a hydroxyl

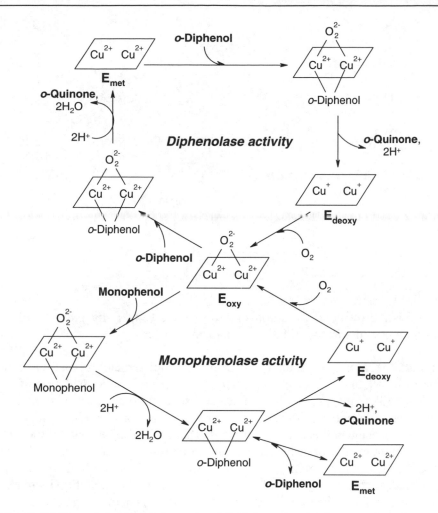

Fig. 3.47. Reaction mechanisms for the diphenolase and monophenolase activities of phenoloxidase

radical ($\cdot$OH) [203, 204], compounds which have the capacity to oxidize the organic xenobiotics. In other words, the formation of these radicals enables the phenoloxidase to participate in detoxification processes by the co-oxidation mechanism presented below (Fig. 3.48).

Analogously, nitrobenzene is oxidized to m-nitrophenol, and the methyl group of [C^3H_3] TNT [2] is oxidized by an enzyme preparation isolated from tea leaves.

Data indicating the participation of plant phenoloxidases in the oxidative degradation of xenobiotics are sparse [515], despite the fact that such activity should definitely be expected. Laccases of fungi have been better ex-

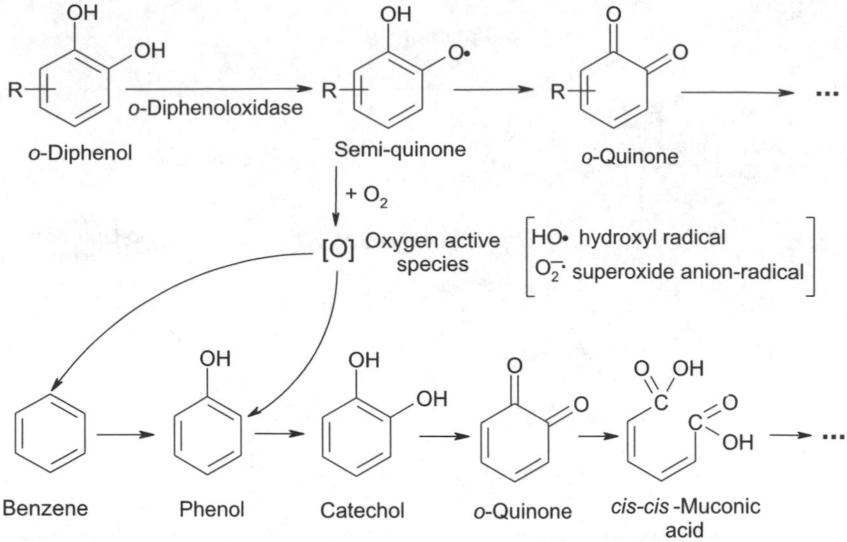

Fig. 3.48. Enzymatic oxidation of *o*-diphenols (upper) by phenoloxidase and non-enzymatic co-oxidation of benzene (lower)

plored. Laccases biodegrade (oxidize) many aliphatic and aromatic hydrocarbons [89], and also actively participate in the enzymatic oxidation of alkenes [357]. Crude preparations of laccase isolated from the white rot fungus *Trametes versicolor* oxidize 3,4-benzopyrene, anthracene, chrysene, phenanthrene, acenaphthene and other PAHs [88, 257]. The intensity of the oxidation of these environmental contaminants increases in the presence of such mediators as phenol, aniline, 4-hydroxybenzoic acid and 4-hydroxybenzyl alcohol, which are substrates of laccase. The rate of PAH oxidation increases proportionally to the redox potential of the mediators in the range $E_h < 0.9$ V. The rate decreases with redox potential in the range $E_h > 0.9$ V [257]. The natural substrates of laccase, methionine and cysteine, reduced glutathione, and others, also stimulate the oxidation of xenobiotics. These data indicate that in the cases of laccase and *o*-diphenoloxidase, the oxidation of hydrocarbons is carried out by a co-oxidation mechanism [508, 511, 515].

Esterases

Several lipophilic organic contaminants containing ester bonds acquire functional groups not only via oxidation, but also via hydrolysis. Among them are such compounds as phthalate esters (chemical plasticizers), 2,4-D, diclofop-methyl, bromoxynil octanoate, binapacryl, aryloxyphenoxypropionate, pyrethrin, methyl paraoxon, malathion (pesticides), etc. The functionalization of organic xenobiotics via hydrolysis is catalyzed by serine hydrolases such as carboxylesterases (EC 3.1.1.1) [95, 96, 281, 428]. These enzymes, being isolated from microsomes, have a wide specificity and in addition to their primary reaction (hydrolysis of carboxyl ester with formation of an alcohol and a carboxyl acid) also catalyze the hydrolytic cleavage of other bonds, viz. arylesterase (EC 3.1.1.2), lysophospholipase (EC 3.1.1.5), acetylesterase (EC 3.1.1.6), acylglycerol lipase (EC 3.1.1.23), acylcarnitine hydrolase (EC 3.1.1.28), palmitoyl-CoA hydrolase (EC 3.1.2.2), amidase (EC 3.5.1.4), aryl-acylamidase (EC 3.5.1.13), etc. A broad specificity allows the esterases to actively participate in the functionalization phase of lipophilic xenobiotics.

One of 12 non-specific esterases in wheat shows a preference for a substrate with chain-length of 6–8 carbon atoms, and this form of esterase is active with the toxic chemical plasticizer *bis*(2-ethylhexyl)phthalate [95]. The cleavage pathway of this molecule is presented in Fig. 3.49.

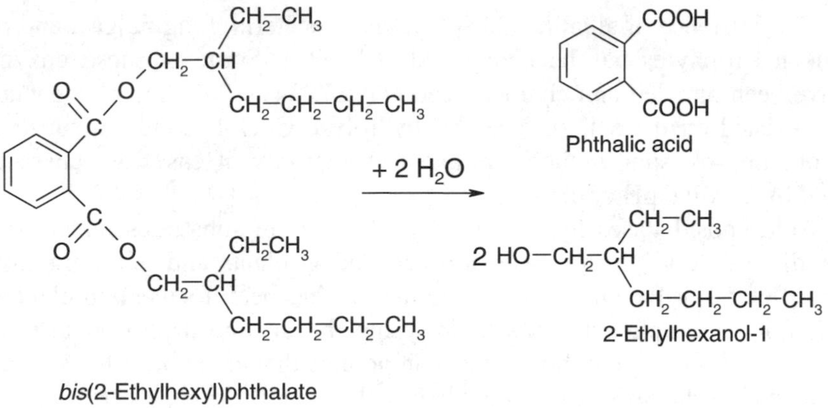

Fig. 3.49. Hydrolysis of *bis*(2-ethylhexyl)phthalate by esterase

Esterases also effectively hydrolyze model xenobiotics such as *p*-nitrophenyl acetate and α-naphthyl acetate. Comparison of various plant esterase activities showed that activity toward model xenobiotics was the highest in wheat, while esterases in weeds (wild oat, black-grass (*Alopecurus myosuroides*)) were more active in pesticide ester hydrolysis (diclofop-methyl, bromoxynil octanoate, binapacryl) [96]. This distinction is simply a consequence of the wide variety of esterases in plants. Weeds contain more basic esterases (pI > 5.0) with a high affinity towards pesticides, while the acidic esterase (pI = 4.6) from wheat has the greatest activity toward α-naphthyl acetate but no activity towards pesticides.

The esterase family in plants is important for the endogenous metabolism and herbicide bioactivation in crops and weeds. Carboxyesterase, a member of the family of serine hydrolases (designated by GDSH), which activate aryloxyphenoxypropionate graminicides towards their bioactive herbicidal acids by hydrolyzing the respective ester precursors have been identified in black-grass, a problem weed of cereal crops in Northern Europe [95]. This enzyme (designated by *Am*GDSH1) was cloned and expressed in yeast (*Pichia pastoris*) as a secreted form of the enzyme. Expression was associated with activity towards aryloxyphenoxypropionate esters. *Am*GDSH1 was predicted to be glycosylated and exported to the apoplast of plants.

A special subclass of esterases is enzymes classified in a unified manner as phosphatases (EC 3.1.3), catalyzing the hydrolysis of ester (P–O–R) and anhydride (P–O–P) bonds by liberating phosphoric acid. According to their action at different pHs the existence of two classes of phosphatases is recognized: alkaline phosphatases (EC 3.1.3.1) and acid phophatases (EC 3.1.3.2).

The existence of alkaline phosphatases in bacteria, fungi, algae, and marine macrophytes has been detected [12, 537]. Some of these enzymes have been purified and characterized [12, 17, 347]. Alkaline phosphatases are located on the cell surface and hydrolyze C–O–P bonds in substrates containing phospho-monoesters and in a majority of cases are characterized by alkaline pH optima.

Acid phosphatases hydrolyze C–O–P bonds in substrates under acidic conditions. Acid phosphatases are detected in plants and microorganisms [64, 189, 537]. Recently, increasing interest has been focused on plant and microorganism phosphatases hydrolyzing the di- and triphosphate esters. An example of organophosphorus compounds that are subject to the action of acid phosphatases is presented below (Fig. 3.50) [537].

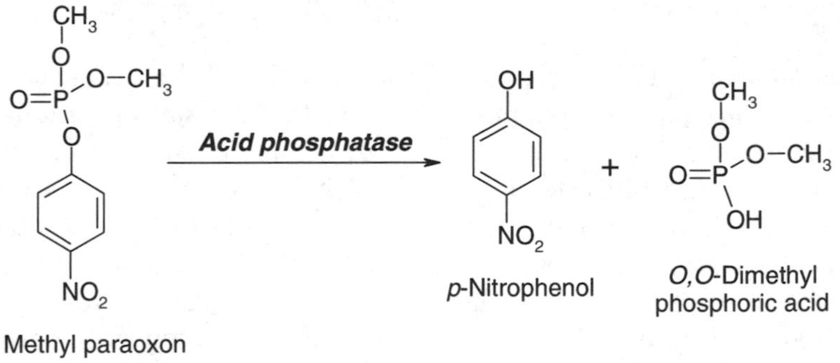

Fig. 3.50. Formation of *p*-nitrophenol from methyl paraoxon under the action of acid phosphatase (EC 3.1.3.2)

It has been shown that phosphatase from giant duckweed (*Spirodela oligorrhiza*), mung bean and slime mold (*Dictyostelium discoideum*) are capable of hydrolyzing such highly toxic organophosphates as the nerve agents phosphofluoridic acid-(1-methylethyl) ester (DFP), soman, and sarin (Fig. 3.51) [236]. It has also been demonstrated that aquatic plants such as parrot feather, giant duckweed, and Canadian waterweed rapidly transform the following organophosphate compounds: diethyl ((dimethoxyphosphinothioyl)thio)butanedioate (malathion), S-(2-(ethylthio)ethyl) O,O-dimethylphosphorothioate (demeton-S-methyl), and 2-chloro-4-(1,1-dimethylethyl)phenylmethyl methylphosphoramidate (crufomate) (Fig. 3.51) [184].

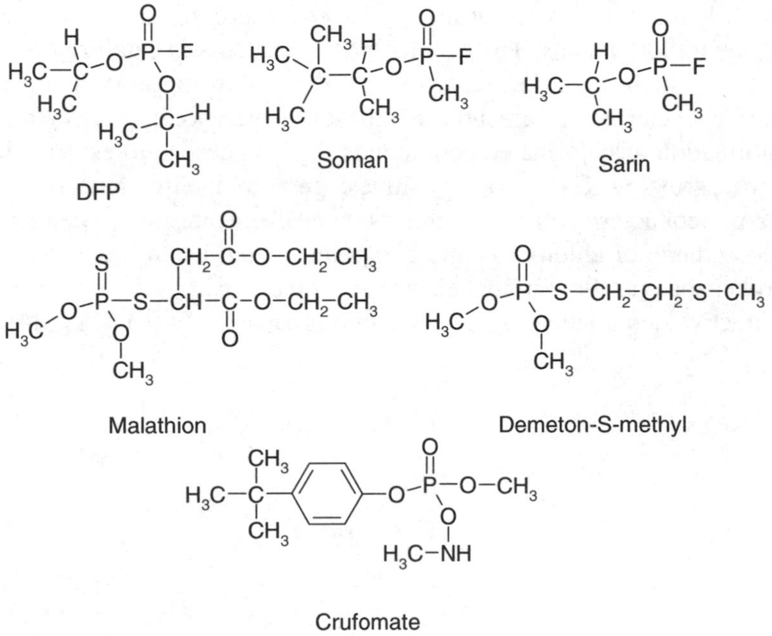

Fig. 3.51. Organophosphate pesticides hydrolyzed by acid phosphatase

There are some other organophosphate compounds that are also hydrolyzed by enzymes, some of them being mixed-function oxidases [40], and flavin-containing monooxygenases [307], catalyzing P=S bond oxidation. In this list, carboxylesterase (EC 3.1.1.1) hydrolyzing the C–O bond should also be mentioned [290].

One should also note another group of enzymes, the organophosphate acid anhydrases, which specifically catalyze organophosphate hydrolysis [78, 543].

Dehalogenases

Halogenated pollutants (dioxins, PCBs, chlorinated solvents, organochlorine pesticides, etc.) are the most widely spread toxic and persistent compounds. Their detoxification involves the removal of halogen atoms (dehalogenation). Dehalogenation generally makes xenobiotic compounds less toxic and more easily degradable [254, 344]. This process occurs via reductive or oxidative reactions (see Fig. 3.34 and 3.35). Reductive dehalogenation occurs primary in both anaerobic and aerobic conditions in microorganisms. The dehalogenation process is catalyzed by dehalogenase, which is a key enzyme in the halorespiratory pathway in anaerobic bacteria that are able to conserve metabolic energy from the dechlorination of chlorinated compounds [516]. These halorespiring bacteria (groups of low G+C Gram-positives, green nonsulfur bacteria, and δ- and ε-proteobacteria) can use chloroalkenes, e.g., tetrachloroethene and trichloroethene or chloroaromatic compounds such as chlorophenols or 3-chlorobenzoate as the terminal electron acceptor.

Reductive dehalogenation involves two processes (Fig. 3.52) [344]:

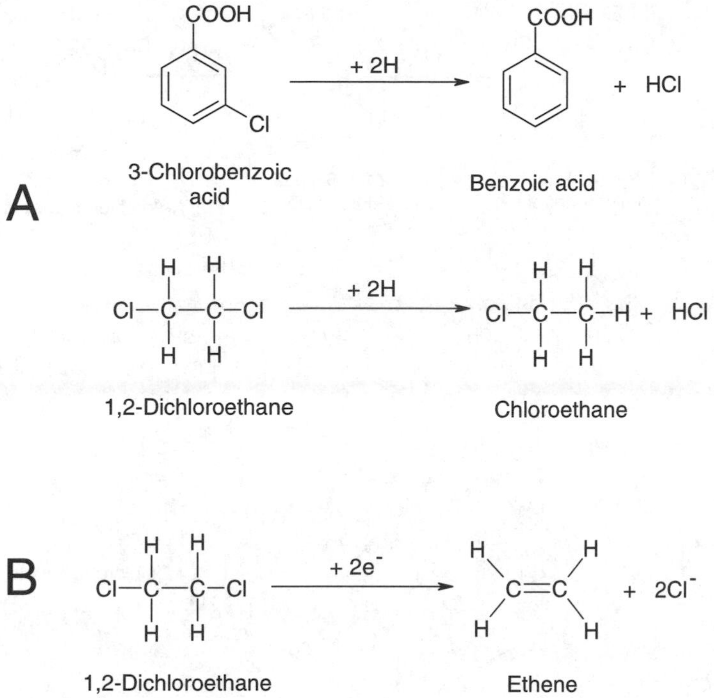

Fig. 3.52. Reductive dehalogenation via hydrogenolysis (A) and vicinal reduction (B) in anaerobic conditions [344]

1. Hydrogenolysis is the replacement of a halogen substituent of alkyl or aryl halogenide molecule with a hydrogen atom (Fig. 3.52A).
2. Vicinal reduction (dihaloelimination), is the removal of two halogen substituents of only alkyl halogenides with the formation of an additional bond between the carbon atoms (Fig. 3.52B)

Both processes require electron donors (reductants) such as NADH, glutathione etc. [254, 537]

Aerobic bacteria (such as *Flavobacterium* sp., *Rhodococcus* sp. etc.) use reductive dehalogenation for the degradation of polychlorinated aromatic rings which are invulnerable to ring-cleaving oxygenases. In aerobic bacteria the dehalogenation process differs from that in anaerobic ones and involves the step of hydrolytic cleavage of the carbon–chlorine bond followed by reductive dechlorinations (Fig. 3.53) [344].

According to the existing data some plant species (soybean, clover, wheat, etc.) are able to dehalogenate polychlorinated pollutants such as PCBs [220]. The specificity of this process depends on quantity of chlorine atoms as substituents and their positions in the aromatic rings.

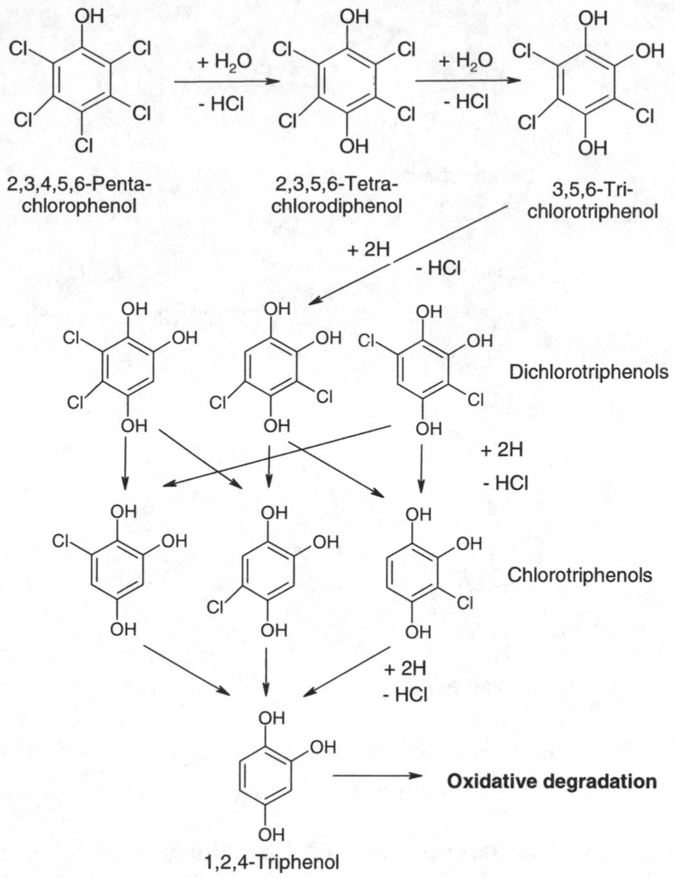

Fig. 3.53. Dehalogenation via hydrolysis and hydrogenolysis in aerobic conditions in *Rhodococcus chlorophenolicus* [344]

Canadian waterweed (*Elodea canadensis*) transforms TCE to dichloro-ethylene and HCl (Fig. 3.54) [537]. This dehalogenation was not observed when using halogenated aromatics as substrates.

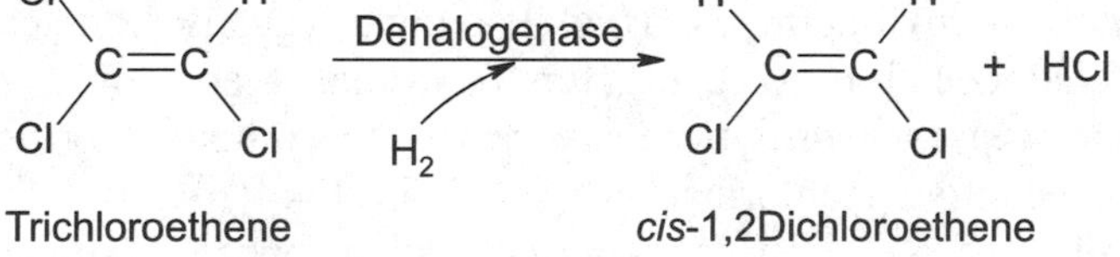

Fig. 3.54. Dehalogenation of trichloroethene to *cis*-1,2-dichloroethene by dehalogenases [537]

Hybrid poplar converts trichloroethylene by another way and forms tri-chloroacetic acid, dichloroacetic acid, and trichloroethanole as metabolites [116, 452]. This fact indicates that TCE may be transformed via an oxidative pathway, possibly involving microsomal cytochrome P450-containing monooxygenase [116].

Nitroreductases

According to the current classification, nitroreductases are enzymes catalyzing the reduction of the nitro groups in aromatic compounds, such as explosives, as TNT, and are classified as EC 1.6.6 non-specific NAD(P)H-dependent nitroreductases [156]. These enzymes are found in animals, plants and microorganisms.

The nitroreductase from *Enterobacter cloacae* contains two monomers and binds two flavin mononucleotide prosthetic groups at the monomer interface [225]. The enzyme procures reducing equivalents from NADH and NADPH by means of two flavin mononucleotide cofactors (FMN). In the oxidized enzyme, the flavin ring system adopts a strongly bent (16°) conformation, and the bend increases (to 25°) in the reduced form of the enzyme, which is roughly the conformation predicted for reduced flavin in solution. Free oxidized flavin has a planar configuration, the induced bend in the oxidized enzyme may favor reduction, and it may also account for the characteristic inability of the enzyme to stabilize one electron-reduced semiquinone flavin, which is planar.

The transformation of TNT is in many respects predetermined by its singular chemical structure. The polarization of the N–O bond, due to the greater electronegativity of oxygen compared to nitrogen, induces a partial positive charge on the nitrogen. This charge, combined with the high electronegativity of nitrogen, makes the nitro group easily reducible. On the other hand, the delocalized π electrons from the aromatic ring of TNT are removed by the electronegative nitro groups, which make the ring electrophilic [387].

To catalyze its reaction, nitroreductase uses reduced pyridine nucleotides (both NADH and NADPH) as electron sources [547]. There are two types of nitroreductases (Fig. 3.55) [156]. Type I is found in animals, plants, and a number of microorganisms (in bacterial genera such as: *Bacillus, Staphylococcus, Pseudomonas*, etc. and actinomycetes). The enzyme reduces the nitro group by two-electron transfers. This pathway is oxygen independent and no radicals are formed [51]. Therefore, the type I nitroreductases expose activity under both anaerobic and aerobic conditions. Type II is oxygen sensitive and reduces nitro groups through single-electron transfers, forming a nitro-anion radical. Under aerobic conditions, an oxygen molecule reacts with a nitro-anion radical and forms a superoxide anion radical that makes the process of TNT transformation reversible. Such nitroreductases are found in rat liver microsomes, and in some strains of *Escherichia coli* [377] and *Clostridium* [11].

The reduction of the first nitro group (in the 2- or 4- position) in TNT is generally much more rapid than that of the remaining groups. The transformation of the nitro to an amino group decreases the electron deficiency of the nitroaromatic ring, and consequently a lower redox potential is required to reduce the rest of the nitro groups from the molecule.

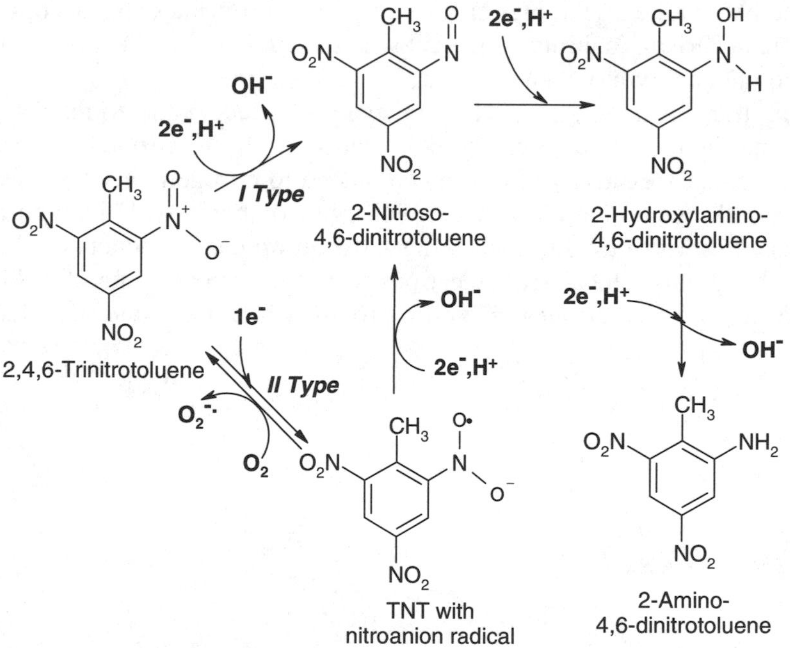

Fig. 3.55. Reduction of TNT by both types of nitroreductases

Nitroreductases catalyze further transformations of the other nitro groups of TNT to amino groups. The removal of the nitro group from the *o*-position and subsequent reduction of the removed nitrite ion by nitrite reductase also takes place. As indicated above, the electron deficiency in the aromatic core of TNT induces nucleophilic attack. The hydride anion from the reduced pyridine nucleotides attacks the aromatic ring, and as a result a non-aromatic structure such as a Meisenheimer σ-complex can be formed [159]. Further, a nitrite anion is released from the Meisenheimer complex with the formation of 2,4-dinitrotoluene (Fig. 3.56). Oxygen is not required for the formation of this compound, and, thus, this process is an alternative for the metabolism of nitroaromatic compounds when oxidative removal of the nitro groups is not possible.

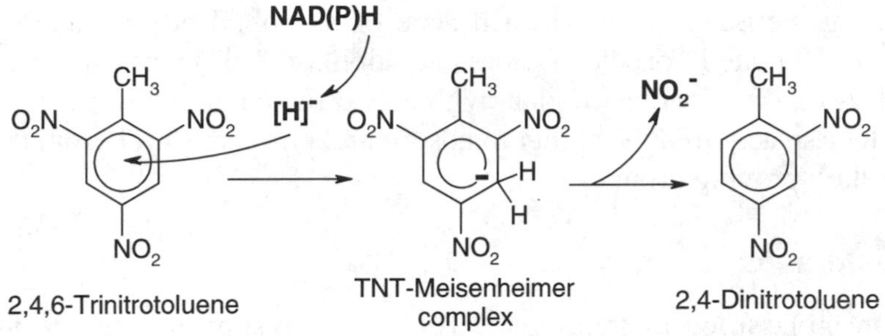

Fig. 3.56. Mechanism of enzymatic elimination of the first nitro group from TNT

The nitro, nitroso and hydroxylamino groups are responsible for the toxicity and mutagenic activity of TNT and its derivatives. It has been shown that complete reduction of the nitro groups to amino groups significantly decreases the mutagenic effect of this contaminant [67].

The presence of a highly active nitroreductase is required for plants that are used to phytoremediate explosive-contaminated soils and groundwater. The correlation between the plant nitroreductase activity and ability to absorb TNT from aqueous solutions has been demonstrated, with the corresponding increase of nitroreductase activity paralleling the faster assimilation of TNT by the plant [267]. These results support the hypothesis that plant nitroreductase activity may serve as a simple preliminary biochemical test to select plants with potential for the phytoremediation of areas contaminated by TNT and similar nitroaromatic compounds. Some plants actively absorb and transform TNT: yellow nutsedge [366], bush bean [222], switch-grass [379], parrot feather, stonewort, algae, ferns, monocotyledonous and dicotyledonous plants, aquatic and wetland species [30], hybrid poplar [497] and soybean [267]. In the transformation of TNT by plants formation of the monoamino derivatives 2-amino-4,6-dinitrotoluene and 4-amino-2,6-dinitrotoluene takes place. A large proportion, sometimes about 60%, of the metabolites seems to get involved in conjugation with insoluble biopolymers [2, 36, 451], often with lignin and hemicellulose [440]. These conjugates are compartmentalized into the vacuoles and the cell wall.

The use of a plant-bacterial consortium (e.g., *Pseudomonas* strain together with meadow bromegrass (*Bromus erectus*) for the phytoremediaton of contaminated soil has also been reported [464]. The bacteria have an active nitroreductase, able to transform TNT into its monoaminodinitrotoluene and diaminonitrotoluene metabolites that facilitate the easy removal of explosives intermediates from the soil by plants.

Transgenic plants with a cloned gene of microbial nitroreductase have also been created for the phytoremediation of TNT-contaminated soils [179]. A transgenic tobacco plant with an expressed gene of a bacterial nitroreductase acquired the ability to absorb and eliminate TNT from the soil of military proving grounds [213].

Transferases

Enzymes classified as transferases (EC 2) are responsible for catalyzing conjugation reactions of parent toxic compounds and the formation of intermediates with endogenous plant cell constituents. The participation of individual enzymes depends on the chemical nature of the intermediates and the existence of the other cell constituents required. The process of conjugation is carried out by glutathione S-transferase (EC 2.8.1.18), O-glucosyltransferase (EC 2.4.1.7), N-glucosyltransferase (EC 2.4.1.71), N-malonyltransferase (EC 2.3.1.114), putrescine N-methyl-transferase (EC 2.1.1.53) [428] etc. Under normal conditions these enzymes participate in cell metabolism, and in the case of organic xenobiotics penetrating into plant cells at appreciable (or indeed high) concentrations they are involved in the conjugation of xenobiotics. All cell constituents bound with xenobiotics in the conjugation processes have a hydrophilic nature, and thus the hydrophobicity of the toxicants decreases as a result of conjugation. Therefore, the conjugates are more soluble in the cytoplasm than the parent toxicants and undergo compartmentalization.

For instance, conjugation with the tripeptide glutathione proceeds according to a widespread detoxification mechanism for xenobiotics in plants and mammals. Conjugation with glutathione has been demonstrated for chemical groups as diverse as chloroacetamides, triazines and sulfonylureas [158, 301] and is regarded as an important contributor to the selectivity of herbicides in target crops.

The group of enzymes called glutathione S-transferases, GSTs (other names encountered in the literature are glutathione S-alkyltransferase, glutathione S-aryltransferase, S-(hydroxyalkyl)glutathione lyase, and glutathione S-arylalkyltransferase) has wide specificity and couples electrophilic xenobiotics and their metabolites with the reduced tripeptide glutathione. In plants, a large and diverse gene family encodes the glutathione transferases. GSTs facilitate the reaction between the functional group of the contaminant intermediates and the SH-group of the glutathione cysteine residue. They participate in the conjugation of a wide spectrum of toxic compounds such as the herbicides flufenacet (FOE 5043), triflusulfuron, chlorimuron-ethyl, acetochlor, metolachlor, alachlor, atrazine [37], safeners [107], fluorodifen [117], etc. In consequence, the toxicant is bound to intracellular compounds via a covalent bond to the sulfur atom (Fig. 3.57).

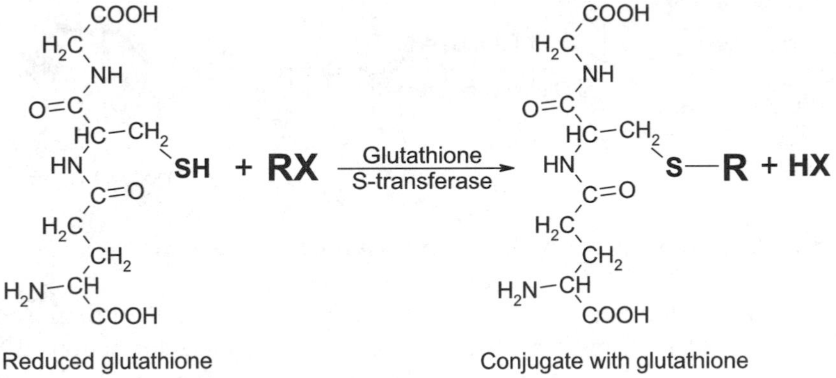

Fig. 3.57. The conjugation reaction with reduced glutathione

In the reaction presented in Fig. 3.57, R is an aliphatic, aromatic, heterocyclic, sulfate, nitrite, or other group. The GSTs also catalyze the addition of aliphatic epoxides and arene oxides to glutathione; the reduction of polyol nitrate by glutathione to polyol and nitrite; certain isomerization reactions and disulfide exchange.

GSTs constitute an important part of the cellular detoxification system and are found not only in plants but also in virtually all organisms that have been investigated. Irrespective of species, the soluble cytosolic GSTs are classified as: α, μ, π, σ, or θ on the basis of their protein sequence and immunologic cross-reaction [301]. So far, all plant GSTs have been assigned to Class θ, a very heterogeneous group of proteins found in various organisms exhibiting differences in the primary sequence [158].

Glucosyltransferases catalyze the reaction between glucose and hydroxyl or amino groups (N-glucosyltransferases) of xenobiotics [318]. Both enzymes are inducible under the action of some herbicides and other organic contaminants (e.g., 3,4-DCA, 4-nitrophenol and 2,4,5-trichlorophenol) [41].

In plants, xenobiotics undergo conjugation with the help of various transferases. For example, the herbicide 3,4-DCA is metabolized via N-malonyltransferase in soybean root cultures, but in *Arabidopsis thaliana* root cultures via N-glucosyltransferase (Fig. 3.58) [291].

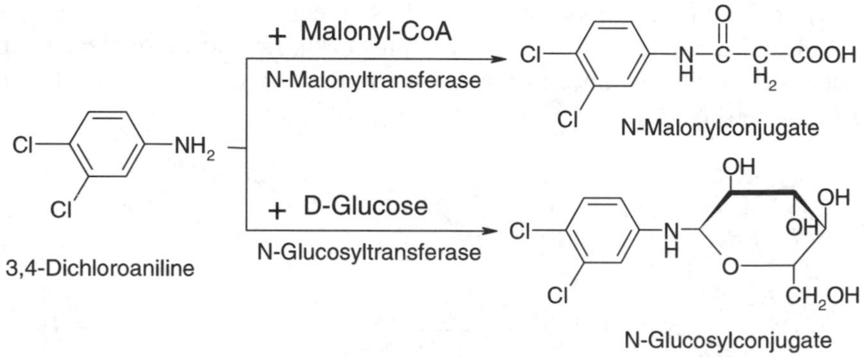

Fig. 3.58. Two pathways of 3,4-dichloroaniline conjugation

2,2-*bis*(4-Chlorophenyl)-acetic acid, the first intermediate of the insecticide DDT metabolism in soybean, is conjugated by the formation of O-glucoside [428]. The conjugation capacity of soybean O-glucosyltransferase is 855 μg 2,2-*bis*(4-chlorophenyl)-acetic acid per hour per gram of cell fresh weight [525].

The best characterized among the plant proteins involved in detoxification are the GSTs from maize. Six isoenzymes differing in their DNA sequence, regulation and substrate specificity have so far been reported. Involvement in the detoxification of herbicides has been demonstrated for the isoenzymes GST I, and GST III, that are constitutively expressed in maize, and for the isoenzymes GST II and GST IV induced in maize after treatment by safener [125].

*   *   *

The material presented above clearly indicates that enzymes and enzymes alone represent the main and only xenobiotic degradation power in all kinds of organisms. Knowledge of the variety of enzymes and levels of their activities is the basis of any kind of phyto- or bioremediation technology.

Undoubtedly determination of the detoxification potential of entire plants and the elucidation of the enzymatic mechanisms of phytoremediation depend both upon the establishment of a sound theoretical basis and the accumulation of more experimental data. However, even the existing level of knowledge allows the creation of qualitatively new phytoremediation technologies for remediation and long-term protection of the environment.

Phytoremediation is a complicated process depending on many factors, such as plant species and variety, the activity of microorganisms, fertility of the soil, type of contaminant, temperature, pH, humidity, oxygen concentration in the soil, etc. That is why it is difficult to determine the real input of each factor in the bioremediation process, but it is absolutely clear that in plants as well as in microbes intracellular enzymatic degradation of contaminants is mainly carried out by oxidative enzymes, namely cytochrome P450-containing monooxygenase, peroxidase, phenoloxidase and others in plants, and lignin peroxidase, Mn-dependent peroxidase, and laccase in microorganisms. Despite the fact that in the process of phytoremediation quite a few enzymes actually participate, we currently do not have a clear understanding of the mechanisms of their action. Particular challenges are to understand their interreplacement during deep oxidation of organic contaminants, the optimal correlation of their activities, their substrate specificity, etc. To overcome limitations in the remediation process, it should be ascertained which enzymatic activity is deficient for a particular plant in order to be able to create a new, genetically modified, effective plant-transformant. Over the past decade various genetic engineering approaches have been carried out to improve the efficiency of plants in removing and degrading contaminants. For this reason, it is extremely important to know which enzymatic reactions (oxidases, reductases, esterases, transferases, etc.) are limiting factors for the remediation process. The degradation pathway of some contaminants appears to be significantly complicated, but in part the complicated appearance is undoubtedly due to the present lack of essential information. This is the main gap in our incomplete understanding of the detoxification process.

4 The ecological importance of plants for contaminated environments

Plants, in combination with microorganisms, have been fashioning the organic world for over a thousand million years through photosynthesis, fixation of molecular nitrogen and metabolic transformations of their environment. As a result of the action of plants, microorganisms, animals (insects, birds, etc.) as well as anthropogenic and abiotic factors, the chemical composition of our planet has been brought to a more or less uniform and steady state. During the last million years or so, human beings also participated in the distribution of organic substances. Later on, due to increased human activities, the types of organic and inorganic compounds in the air, soil and water reservoirs became less desirable. The creation and distribution of these technogenic chemicals became especially obvious in the nineteeth and twentieth centuries, as a result of military action, relentlessly increasing manufactures, increased oil production, transportation, use of pesticides, etc. Terms such as ecocontaminant, ecotoxicant and ecopollutant date from a comparatively recent time. In the new ecological situation, typified by massive emissions from chemical and other industrial factories, plants exhibited new capabilities for the absorption and metabolic degradation of technogenic pollutants. In fact, the creative or generative activity of plants includes not only the formation of vitally important organic compounds, but also the maintenance of their environment, especially control of the levels of toxic compounds. Hence, they may eliminate or at least significantly decrease ecological hazards, depending on their genetically determined abilities.

In order to properly understand the ecological power of plants and evaluate their detoxification potential, the anatomical-morphological and physiological-biochemical particularities of plants responsible for establishing the basis for their role as environmental protectors and remediators should be emphasized:
- Higher plants are in contact with soil and water through roots and with the air by leaves, and thus interact simultaneously with three different environments.

– Soil-plant-microbial interactions engender unique processes influencing the overall plant metabolism as well as transformations of xenobiotics.
– Highly developed root systems allow plants to control large areas of soil at different depths and create micro-conditions convenient for the multiplication of microorganisms in their rhizosphere with the help of exudates.
– The large surfaces of plant leaves permit the absorption of pollutants from the air via the cuticle (lipophilic compounds) and stomata (gases).
– A well-developed internal transportation system for nutrients works in both directions and allows environmental contaminants to be distributed throughout the entire plant.
– Plant-microbial interaction creates a microenvironment resulting in the concentration and penetration of contaminants at and into the roots.
– The autonomous synthesis of vitally important organic compounds by photosynthesis requires primary ammonia (via uptake of nitrate or ammonia from the soil, or in the case of leguminous plants as a result of symbiotic nitrogen fixation). This process is important since remediation of polluted sites requires additional metabolic force especially in the case of prolonged contact with contaminants.
– Plants contain the apparatus required for the full set of biochemical and physiological processes of detoxification, and have no need for additional non-plant microorganism-based technological help.

One should moreover note:

– The existence of constitutive and inducible enzymes catalyzing degradation, conjugation and other detoxification processes.
– The availability of a large intracellular space to deposit the residues of toxic compounds as conjugates, and to accumulate heavy metals.
– The functionalization or further transformation of organic contaminants in plant cells (conjugation, deep oxidation etc.), depending on the molecular structure of the contaminant.

These characteristics confer advantages distinguishing higher plants from other organisms and determining their universal ability to absorb the great majority of contaminants from soil, water and air. To evaluate the ecological potential of higher plants one should keep in mind that in spite of global urbanization plants still occupy more than 40% of the land.

The main disadvantage of plant-based cleaning technologies is their strong dependence on climate. Climate is a very important factor for growth, development and metabolic activity of plants; climate is the main limiting factor in the distribution and survival of plants. The decrease of plant detoxification activity is especially noticeable in areas remote from the equator. According to their tolerance of temperature, plants are divided into the following groups: equatorial, tropical, subtropical, Mediterranean, warm-temperate, cool-temperate, boreal, and polar [521]. For instance, equatorial evergreen tropical rainforest plants permanently maintain ecological activity regardless of the season. Other plants, even subtropical ones, are influenced by the seasons. Boreal and treeless tundra vegetation types with a short or very short summer and a long and cold winter are less effective for environment decontamination. All above-mentioned plant groups have been well characterized according to their ecological importance [292]. Microclimate, determined by zonal and regional factors, is an important parameter for the cultivation of plants. The zonal climate is the result of the energy balance prevailing at different latitudes. Regional climate depends on the distance from seas, oceans, ocean currents (warm: e.g., Gulf Stream, Kuroshio; cold e.g., Labrador, Benguela), and such factors as prevailing winds (and position with respect to mountain barriers, e.g. leeward), etc. Climate and weather are continuously changing albeit at different time scales. The interaction of meteorological factors, such as air temperature, humidity, wind speed and direction, and precipitation (its quality and nature) with each other, determines the frequency of fluctuations in weather, which, in turn, significantly affects plant detoxification abilities.

The activity of the rhizospheric microflora also depends on temperature. Temperature determines the substrate exchange intensity between soil, plant root systems and microbial consortia [292]. As a result of contacts between these compartments, the root system takes up inorganic compounds dissolved in water and nutrients and releases exudates (enzymes, carbohydrates, vitamins, organic acids, growth factors, etc.). The exudates in the soil create a favorable environment for the development of microorganisms of different taxonomic groups (bacteria, actinomycetes, fungi). The development of soil microorganisms and their intimate contact with root systems leads to the activation of potential metabolic interactions between roots and microorganisms. The mucus covering of the root caps provides an additional substrate for soil microflora. The profit for plants in this environment is the acceleration of exchange processes and the increased velocity of nutrient uptake and transportation, including organic substrates and heavy metals.

A high potential to carry out biochemical reactions, and an increased absorption surface developed in the root systems, allow mycorrhizal fungi to participate in nutrient uptake [292]. The worldwide distribution of these fungi indicates their supreme importance in plant-microbial interactions. These fungi belong to the subclass of lower fungi – Zygomycetes. Most plants are hosts of mycorrhizal fungi. Interaction of these fungi with plants is highly mutually desirable. Plants provide the fungi with carbohydrates and receive back chemical elements and organic compounds required for the plants' growth and activation of root metabolic processes.

The plant cell is not a small factory permanently able to absorb and metabolize organic contaminants of different structures. Oxidative or any other kind of xenobiotic degradation (reduction, hydrolysis) requires extra energy, which a plant cell has to provide. There are intracellular processes in plants that undergo deviations from the norm as a result of the penetration of xenobiotics into plant cells. Plant cell ultrastructure is sensitive to the action of contaminants, and according to the changes engendered by them, their intracellular concentrations can be classified as metabolizable or lethal [544]. The transformation of contaminants is closely related to the metabolic activity of a plant cell. The present authors have shown (hitherto unpublished data) that small doses of contaminants with aliphatic and aromatic structures induce the activation of key Krebs cycle enzymes such as malate dehydrogenase and enzymes providing the plant cell with nitrogen-containing organic compounds, including catabolic fuel in the case of energy deficiency (glutamine synthetase, glutamate dehydrogenase) [288]. The extra energy required by detoxification processes is partially spent on the induced synthesis of the enzymes participating in xenobiotic degradation, their movement and deposition, e.g. in vacuoles. Detoxification processes are connected to photosynthesis, the intensity of which is significantly decreased under the influence of contaminants [182, 454]. Xenobiotic transformations are closely related to the majority of intracellular metabolic processes requiring extra energy, and this energy dependence is one of the main limiting factors in the detoxification potential of plants.

A plant's ecological potential is directed to remove contaminants from the environment. The purposeful application of plants can be of long-term or of short-term advantage, depending on the targeted goal. For phytoremediation, long-term application of the planted system to exploit and amortize its potential and maintain its continued effectiveness is recommended. Monitoring should follow short-term cleanup. Essentials of monitoring whose results are used for plant selection include the following: type and concentration of constituent elements; frequency and duration; sampling methods; locations; and quality control requirements [152]. Sometimes, a phytoremediation process would take too much time and, hence, would not be acceptable. All phytoremediation technologies depend on the growth rates of the plants, which are characterized by seasonal activity; growth rate often limit the application efficiency. Several growing seasons may be needed to attain the effective age of the plant for optimal phytoremediation.

4.1 Plants for phytoremediation

The realization of phytoremediation technologies implies the planting of a contaminated area with one or more specific, previously selected plant species with the potential to extract contaminants from the soil. The treatment continues by harvesting the plants, composting, disposal in a landfill, or incinerating them. To create a truly effective phytoremediation system all components of the system should be thoroughly analysed. The major constitutive component of such a system is obviously the plant itself. The goal of plant selection is to choose a plant species with appropriate characteristics.[1] A survey of the vegetation on site should be undertaken to determine what species of plants would have the best growth at the contaminated site, taking into account the abilities of the plants to accumulate and degrade the contaminants.

The assessment of the detoxification potential of the plant is determined by the rate and depth of contaminant uptake from the soil, accumulation in the plant cell, and the degree of contaminant transformation to regular cell metabolites. The best plants for a particular phytoremediation task should be selected based on multiple plant characteristics. First, the actual phytoremediation–related characteristics of the candidate plants should be established, notably:
– Overall ability to take up and degrade contaminants in the soil or groundwater.
– Ability to accumulate organic and inorganic contaminants in their cells and intracellular spaces.
– Excretion of exudates to stimulate the multiplication of soil microorganisms and secretion of enzymes participating in the initial transformations of the contaminants.
– Existence within the cells of contaminant-degrading or conjugating enzymes (oxidases, reductases, transferases, esterases, etc.).

---

[1] At present, very little work has been done on increasing the effectiveness of the phytoremediation system by creating biocoenoses, or communities, in which the plants act synergistically. In other words, at present the focus is on single contaminants, and the goal of selection is to find specific plant species for their elimination. A considerable body of knowledge, acquired from laboratory and field experiments, exist for identification of the optimal candidate. On the other hand, where several contaminants need to be eliminated, a corresponding mixture of the individually optimum candidates may not constitute the best solution. To date, only some types of synergisms between plants and microorganisms have been investigated.

- High resistance against contaminants, i.e., that the plants' growth and metabolism is not adversely affected by the contaminants.
- The root system (main and fibrous); the range of rooting depth of the plants.
- Whether the plants are endemic and non-agricultural.
- Tolerance to salty soil (halophilicity).
- Appropriate adaptation to warm or cold conditions.
- Growth rate.

It is assumed that the nature and level of the contaminants have already been determined. It is also important to establish the localization and distribution of toxic compounds (area and depth of contamination). Important environmental factors bearing on the selection of the best remediation technology are: soil type and characteristic parameters (pH, average humidity, salt content, metal concentration), the presence of parasites, and the expected amount of precipitation during the duration of the remediation process.

Table 4.1 below presents plants widely used in phytoremediation.

It is generally true that the planting of almost any kind of vegetation, including agricultural flora, is beneficial for the human environment. However, in order to make the most of the ecological potential of each particular plant, the selection should be carried out according to the listed criteria. Undoubtedly, technologically the most important part of the plants is the root system, which takes up contaminants from the soil and performs the initial stages of their transformation or accumulation within it. Therefore it is clear that the type of roots and their depth, distribution and type, and degree of ramification are extremely significant components for the successful realization of any phytoremediation technology. A so-called fibrous root system has numerous fine roots providing maximum contact with soil due to the high surface area of the roots. The rooting depth of plants greatly differs between individuals and species (Table 4.2).

Some plants are able to accumulate metals, but the low growth rates typical of these plants limit the total biomass and indicate that the total mass of accumulated metals will be low. Better extraction of toxic compounds from soil may be achieved by the use of mixed plant cultures, but at present there are very little data on their effectiveness.

Table 4.1. Promising plant species for the remediation of sites polluted by organic contaminants

Organic contaminant	Plant Species	Comments	Refs
1	2	3	4
Aromatic hydrocarbons (benzene, toluene)	Maple (*Acer campestre*) Oleaster (*Elaeagnus angustifolia*) Locust (*Robinia pseudoacacia*) Caucasian pear (*Pyrus caucasica*) Walnut (*Juglans regia*) Almond (*Prunus amygdalus*) Cherry (*Cerasus avium*) Cherry (*Cerasus vulgaris*) Amorpha (*Amorpha fructicosa*) Chestnut (*Castanea sativa*) Apple (*Malus domestica*) Zelkova (*Zelkova caprinifolia*) Poplar (*Populus canadensis*) Ryegrass (*Lolium perenne*) Lilac (*Syringa vulgaris*) Weeping willow (*Salix*) Catalpa (*Catalpa bignonioides*) Oriental plane (*Platanus orientalis*) Sophora (*Sophora japonica*)	Plants capable of absorbing 1-10 mg of benzene and toluene per kg fresh leaves per day from air	[127, 510, 515]
	Alfalfa (*Medicago sativa* L.)	Can remove benzene from soil	[165]
		Can enhance biodegradation of toluene by associated microorganisms	[101]

Table 4.1. (cont.)

1	2	3	4
Gaseous alkanes (methane, ethane, propane, butane)	Tea (*Thea sinensis*) Vine (*Vitis vinifera*) Poplar (*Populus canadensis*) Walnut (*Juglans regia*) Maple (*Acer campestre*) Ryegrass (*Lolium multiflorum*) Maize (*Zea mays*) Kidney bean (*Phaseolus vulgaris*)	Plants capable of absorbing 0.1-10 mg of gaseous alkanes per kg fresh leaves per day from air	[508]
	Pine (*Pinus sylvestris* L.)	Plant roots enhance rhizospheric degradation of PHC in soil	[227]
	Alfalfa (*Medicago sativa* L.)	Can remediate crude oil-contaminated soil	[535]
Petroleum hydrocarbons (PHC)	*Spartina alterniflora* (salt marsh sp.) *Juncus roemerianus* (salt marsh sp.) *Spartina patens* (brackish marsh sp.) *Sagittaria lancifolia* (fresh marsh sp.) Clover (*Trifolium* sp.)	Can remediate oil spills in marshes	[311]
	Tall fescue (*Festuca arundinacea* Schreber) Bermuda grass (*Cynodon dactylon*) Ryegrass (*Lolium multiflorum*)		[446]
	Ryegrass (*Lolium perenne*)	Can remediate PHC-contaminated soil and dredged material	[33, 237]

Table 4.1. (cont.)

1	2	3	4
	Ryegrass (*Lolium multiflorum*) Hybrid poplar (*Populus* sp.) Clover (*Trifolium* sp.)		[488]
	Sorghum (*Sorghum bicolor*) Switch grass (*Panicum virgatum*)	Enhance rhizospheric degradation of PAHs in soil	[401]
PAHs	Big bluestem (*Andropogon gerardii*) Little bluestem (*Schizachyrium scoparius*) Indian grass (*Sorghastrum nutans*) Switch-grass (*Panicum virgatum*) Canadian wild rye (*Elymus canadensis*) Western wheatgrass (*Agropyron smithii*) Side oats grama (*Bouteloua curipendula*) Blue grama (*Bouteloua gracilis*)	A mixture of prairie grasses that degrade PAHs	[13]
	Tall fescue (*Festuca arundinacea* Schreber) Alfalfa (*Medicago sativa* L.)	Plants capable of absorbing and degrading naphthalene	[446]
	Prairie buffalo grass (*Buchloe dactyloides*) Kleingrass (*Panicum coloratum* var. 'Verde')	Can decrease naphthalene content in clay soil	[392]
Nitrobenzene	Soybean (*Glycine max* L. Merr. cv. Fiskby v)		[170]
Phenols	Soybean (*Glycine max* L. Merr. cv. Fiskby v) Cane (*Scirpus lacustris* L.)		[449]

Table 4.1. (cont.)

1	2	3	4
Phenols (cont.)	Alfalfa (*Medicago sativa* L.)	Can enhance degradation of phenol by associated microorganisms	[165]
	Potato (*Solanum tuberosum*) White radish (*Raphanus sativus*) Horseradish (*Armoracia rusticana* P. Gaerter, Meyer & Schreb)	Plants with a highly active peroxidase that oxidizes phenols (used in waste-water treatment)	[104]
Polychlorinated solvents	Hybrid poplar (*Populus trichocarpa x P. deltoides*), Aspen (*Populus* sp.) Cottonwood (*Populus* sp.)	Poplars that transpire, metabolize or mineralize 98% of TCE in soil at a concentration of 260 mg per kg	[264]
	Soil green alga (*Chlamydomonas reinhardtii*) Marine green alga (*Dunaliella tertiolecta*)	Algae capable of absorbing and degrading TCE at a concentration of 500 mg per kg soil	[124]
Trichloroethene (TCE)	Wild carrot (*Daucus carota*) Spinach (*Spinacia oleracea*) Tomato (*Lycopersicon esculentum*)	Plants able to absorb and transform TCE from groundwater	[436]
	Waterweed (*Eichhornia crassipes*)		[415]
	Black locust (*Robinia pseudoacacia*)	Can volatilize TCE from groundwater	[355]
Tetrachloroethane	Hybrid poplar (*Populus* sp.)		[61]
	Alfalfa (*Medicago sativa*)	Plant exudates promote degradation of TCE in its rhizosphere	[352]

Table 4.1. (cont.)

1	2	3	4
Polychlorinated solvent (cont.) Dibromoethane TCE	Lespedeza (*Lespedeza cuneata* (Dumont)) Loblolly pine (*Pinus taeda* L.) Soybean (*Glycine max* L. Merr. cv. Davis)	Plants able to increase mineralization of TCE in soil	[7]
	Koa haole (*Leucaena leucocephala* var. K636)	A tropical leguminous tree	[122]
PCBs	Red mulberry (*Morus rubra* L.) Crabapple (*Malus fusca* (Raf.) Schneid) Osage orange (*Maclura pomifera* (Raf.) Schneid)	Plants that produce exudates that stimulate growth of PCB-degrading bacteria	[169]
	Spearmint (*Mentha spicata*)	Plant that induces cometabolism of PCB in its rhizosphere	[191]
	Barley (*Hordeum vulgare* L. cv.Klages)		[333]
Aroclor 1260	Tall fescue (*Festuca arundinacea* Schreb.) Alfalfa (*Medicago sativa* L.) Flatpea (*Lathyrus sylvestris* L.) Lespedeza (*Lespedeza cuneata* Dum) Deertongue (*Panicum clandestinum* L.)	Can enhance mineralization in soil microcosm; decreased levels from 100 to 23-33 mg PCB kg^{-1} soil in 4 months	[154]
Delor 103	Black nightshade (*Solanum nigrum* L.)	Plant effecting 40% mineralization of 100 mg PCB per kg soil in 30 days	[324]
Chlorinated benzoic acid	Slender wheatgrass (*Agropyron pinnata*) Western wheatgrass (*Agropyron smithii*)	Prairie grass species	[463, 500]

Table 4.1. (cont.)

1	2	3	4
Pesticides			
Simazine	Parrot feather (*Myriophyllum aquaticum* (Vell.) Verdc.)		[272]
	Canna (*Canna* x *hybrida* L. 'Yellow King Humbert)		
	Hybrid poplar (*Populus deltoides x nigra* DN34 Imperial Carolina)		[61]
Atrazine	Kochia (*Kochia scoparia* L.Schrad)	Can enhance rhizospheric mineralization of atrazine in soil	[374]
	Pine (*Pinus ponderosa*)	Can support degradation by ectomycorrhizal fungus *Hebeloma crustuliniforme*	[187]
Chloroacetamides	Maize (*Zea mays* L.)		[231]
2,4-D, DDT	Hybrid poplar (*Populus sp.*)		[501]
Metolachlor with Atrazine	Coontail (*Ceratophyllum demersum*) Canadian pondweed (*Elodea canadensis*) Common duckweed (*Lemna minor*)	Aquatic plants that remediate herbicides in water	[403]
Hexachlorobenzene, entachlorobenzene, trichlorobenzene	Barley (*Hordeum vulgare* L. cv.Klages)		[333]

Table 4.1. (cont.)

1	2	3	4
Chlorinated phenols (4-chlorophenol to pentachlorophenol)	Duckweed (*Lemna gibba*)	A floating plant that removes herbicides from water	[148]
Cyanazine with Fluometuron	Ryegrass (*Lolium multiforum* L.) Hairy vetch (*Vicia villosa* Roth) Rice (*Oryza sativa* L.)	Can enhance degradation of herbicides in soil via stimulation of rhizosphere bacterial populations	[520]
Atrazine, Metolachlor, Trifluralin	Kochia (*Kochia scoparia* L.Schrad) Knotweed (*Oiktgibyn* sp.) Crabgrass (*Digitaria* sp.)	Can enhance microbial degradation in rhizosphere, i.e., 45% of atrazine, 50% of metolachlor and 70% of trifluralin in 14 days	[8]
Pentachlorophenol (PCP), Parathion, Diazinon	Wheat grass (*Agropyron cristatum*)	Plant roots enhance rhizospheric degradation of PCP in soil	[164]
	Hard fescue (*Festuca ovina* var. *duriuscula*) Tall fescue (*Festuca arundinacea*) Red fescue (*Festuca rubra*)	A mixture of fescues with high germination rates and high biomass formation in PCP- and PAH-contaminated soil	[385]
	Waterweed (*Eichhornia crassipes*)		[415]
	Crested wheatgrass (*Agropyron desertorum* Fischer ex Link Schultes)	Enhances mineralization of PCP from 100 to 23.1 mg per kg soil in 20 weeks	[164]
	Kidney bean (*Phaseolus vulgaris* cv. 'Tender Green')	Enhances rhizospheric degradation of herbicides	[241]

Table 4.1. (cont.)

1	2	3	4
Bentazon	Black Willow (*Salix alba*) Bald cypress (*Taxodium distichum*) River bitch (*Betula nigra*) Cherry-bark oak (*Quercus falcata*) Live oak (*Quercus virginiana*)	These plants have a high capacity to degrade bentazon	[91]
Aldrin, Dieldrin	Arctic hairgrass (*Deschampsia bernigensis*) Felt-leaf willow (*Salix alaxensis*) Red fescue (*Festuca rubra*) Spikerush (*Eleocharis palustris*)		[532]
Explosives	Switch grass (*Panicum virgatum*)	Prairie grass species	[379]
	Spinach (*Spinacia oleracea*)		[335]
	Parrot feather (*Myriophyllum aquaticum*) Water milfoil (*Myriophyllum spicatum*)	Aquatic plants	[335, 369, 517]
	Stonewort (*Nitella* sp.)	Algae	
2,4,6-Trinitrotoluene (TNT)	Parrot feather (*Myriophyllum aquaticum*) Sweet-flag (*Acorus calamus* L.) Wool-grass (*Scirpus cyperinus* L. Kunth) Waterweed (*Elodea canadensis* Rich. in Michx) Sago pondweed (*Potamogeton pectinatus* L.) Water star-grass (*Heteranthera dubia* Jacq. MacM) Curlyleaf pondweed (*Potamogeton crispus* L.)	Emergent and submerged plant species with a high ability to remove TNT from water and recommended for phytoremediation of explosives-contaminated water around army ammunition plants	[31, 32]

Table 4.1. (cont.)

1	2	3	4
	Kidney bean (*Phaseolus vulgaris* cv. 'Tender Green')		[222]
	Hybrid poplar (*Populus* sp.)		[61, 222]
	Bromegrass (*Bromus erectus*)	A plant-microbial consortium with *Pseudomonas* sp.	[464]
TNT (cont.)	Soybean (*Glycine max*) Ryegrass (*Lolium multiflorum*) Pea (*Pisum sativum*) Chickpea (*Cicer arietinum*)	Plants that can absorb 0.15–0.20 mg TNT per gram fresh biomass per day	[268]
	Hybrid willow (*Salix EW-13*) Hybrid willow (*Salix EW-20*) Hybrid poplar (*Populus ZP-007*) Birch (*Betula pendula*) Norway spruce (*Picea abies*) Pine (*Pinus sylvestris*)	Five-year-old trees of these species are able to uptake TNT from contaminated soil: *Salix EW-20*, 8.5; *Salix EW-13*, 6.0; *Betula pendula*, 5.2; *Populus ZP-007*, 4.2; *Picea abies* – 1.9; *Pinus sylvestris*, 0.8 (g of TNT per m^2 soil per year)	[439]
Hexahydro-1,3,5-trinitro-1,3,5-triazine (RDX)	Parrot feather (*Myriophyllum aquaticum* Vell. Verdc.) Sweet-flag (*Acorus calamus* L.) Wool grass (*Scirpus cyperinus* L. Kunth)		[31, 32]

Table 4.1. (continued)

1	2	3	4
RDX (cont.)	Waterweed (*Elodea canadensis* Rich. in Michx) Sago pondweed (*Potamogeton pectinatus* L.) Water star-grass (*Heteranthera dubia* Jacq. MacM) Curlyleaf pondweed (*Potamogeton crispus* L.)		[31, 32]
	Parrot feather (*Myriophyllum aquaticum*) Spinach (*Spinacia oleracea*) Indian mustard (*Brassica juncea*)		[335]
	Hybrid poplar (*Populus deltoides x nigra*, DN34)		[541]
	Kidney bean (*Phaseolus vulgaris*)		[222]
Octahydro-1,3,5,7-tetranitro-1,3,5,7-tetrazocine (HMX)	Alfalfa (*Medicago sativa*) Kidney bean (*Phaseolus vulgaris*) Canola (*Brassica napa*) Wheat (*Triticum aestivum*) Ryegrass (*Lolium perenne*) Wild bergamot (*Monarda fistulosa*) Western wheatgrass (*Agropyron smithii*) Bromegrass (*Bromus sitchensis*) Koeleria (*Koeleria gracilis*) Goldenrod (*Solidago* sp.) Blueberry (*Vaccinium* sp.) Anemone (*Anemone* sp.) Common thistle (*Cirsium vulgare*) Wax-berry (*Symphoricarpos albus*) Western sage (*Artemisia gnaphalodes*) Drummond's milk vetch (*Astragalus drummondii*)	Terrestrial indigenous and crop plants able to absorb, translocate and accumulate HMX in foliar tissues (from contaminated soil from an anti-tank firing range)	[201]

Table 4.2. Rooting depth of plants used in phytoremediation technologies [151]

Plant	Root depth/m
Indian mustard	0.30
Grasses	0.60–1.20
Alfalfa	1.20–1.80
Poplar trees	4.50

Effective monitoring of the phytoremediation process requires the collection of extensive information, not in the least because many factors influence it apart from the plant species: availability of nutrients, daily maximum, minimum and average temperatures, illumination level (spectral characteristics and irradiance), humidity and its variation, etc. All these parameters should ideally be monitored.

Rapid formation of a large biomass, well-developed roots, and a strong defense system are the most important overall criteria for plants to be successfully applied to the phytoremediation of soils contaminated with heavy metals and organic contaminants.

Is it possible to improve the process of remediation carried out by plants by nongenetical interference? In this regard, encouraging results have been obtained by the application of bioactive preparations.[2] These preparations typically comprise a complex of amino acids and other nutrients including microelements, and are used for plant regeneration following damage due to unfavorable conditions in the surrounding environment, as well as for increasing yield, the development of above-ground biomass, promotion of the assimilation of minerals, facilitation of their transport with the plant sap stream, balancing metabolism, enhancing the plant defense system, metabolic regulation, and are boosters and stress relievers. Some of their characteristics are given in Table 4.3.

According to the authors' recent investigations, these bioactive preparations also increase plant resistance to the action of organic contaminants and heavy metals.

In plants exposed to TNT, benzene and 3,4-benzopyrene the addition of Fosnutren and Humiforte enabled the maintenance of chlorophyll content at the control level, the activation of enzymes participating in the oxidative degradation of pollutants: monooxygenase, peroxidase, phenoloxidase; and the activation of key metabolic enzymes: glutamine synthetase, glutamate

---

[2] For example those from Inagrosa Industrias Agrobiologicas S.A. (Spain) http://www.inagrosa.es/menu productos i.html.

and malate dehydrogenases, i.e. enzymes providing plant cells with nitrogen-containing compounds and energy. Fosnutren and Humiforte contributed to doubling the lead accumulation in roots of maize, kidney bean and yygrass. In shoots the effect of these bioactive preparations on heavy metal accumulation was not noticeable.

Table 4.3. Characteristics of some commercially available bioactive preparations

Product	Shelf life of product sealed in bottle/ years	Dosage of application (Min-Max)/ L ha^{-1}	Mean total quantity used per growth cycle/ L ha^{-1}	Effect and consequences
Aminolforte	4	0.75–1.50	3-4	Increase in yield (20–30%). Reduction in pesticides and fertilizers needed (20–25%)
Fosnutren	6	0.75–1.50	3	Increase in yield (25–30%). Reduction of pesticides and fertilizers needed (20–25%)
Kadostim	5	0.75–1.50	3	Increase in carbohydrates, oils and protein content (30%). Reduction in pesticides and fertilizers needed (20–25%)
Humiforte	6	1.00–2.00	5	Increase in yield in horticulture (up to 50%), elimination of stress

   Not all aspects of the action of bioactive preparations on the phytoremediation processes on differently polluted sites have been investigated yet, but it can be already stated that the preparations typically have a definitely positive effect on plant metabolic activity and promote increased resistance against the action of toxic compounds.

4.2 Phytoremediation technologies

Phytoremediation is a concept constructed from an emerging set of natural technologies to support clean-up strategies. This term is relatively new [397] and means plant-based action (*phyto* – plant, *remediation* – to recover). According to the most modern understanding of phytoremediation technologies, microorganisms also participate as important auxiliaries. Phytoremediation has received special attention in the last decade as an innovative, cost-effective and alternative combination of technological approaches. The main objective of scientists, agronomists, and engineers dealing with phytoremediation is to exploit by the most rational way possible the potential of this natural process. From the technological point of view phytoremediation is the use of vegetation to decontaminate soils and water from heavy metals and toxic organics. Very often, phytoremediation assumes the joint action of both plants and microorganisms.

The most effective technologies based on phytoremediation principles are targeted towards the gradual elimination from the soil of shallow contamination by organics, radionuclides and heavy metals. It is also important to underline that these *in situ* technologies do not damage soil structure and just slightly change soil microbial consortia, even when microorganisms of any taxonomic group (or groups) are introduced into the soil. When comparing the actions of plants and microorganisms on soil it should be noted that in the majority of cases their joint action directed towards xenobiotic degradation exceeds the arithmetical sum of the activity of each individually. At the same time some similarity exists between plants and microorganisms in assimilation of contaminants. After penetration into the cells of both types of organisms contaminants mainly undergo oxidative transformations. Microorganisms, due to their fast growing ability, much more easily regulated adaptation, fast inductive processes and their wide spectrum of enzymes participating in the degradation of organic xenobiotics, are much more active detoxifiers when expressing their activity per unit of dry biomass. A different situation is evident in the case of uptake of inorganics. Microorganisms are often able to accumulate high concentrations of heavy metals. After lysis (degradation) of the microbial cells their content, in spite of the metals, valency change, again becomes a constituent of the soil and hence no remediation takes place. Plants are also able to accumulate high amounts of heavy metals in their roots and subsequently transport them to organs above the ground, eliminating in this way contaminants from the soil.

Plants have shown the capacity to withstand relatively high concentrations of organic contaminants without visually apparent toxic effects. Resistance of plants against any particular contaminant is provided by the plant's cell enzymes being able to degrade it. To select appropriate plants for phytoremediation purposes, they should be thoroughly characterized with regard to their phytoremediation ability before planting. Plants may differ from each other by as many as four orders of magnitude, depending on their ability to uptake and transform contaminants.

Several strategies to use plants for phytoremediation purposes exist. One of the known remediation strategies is planting the contaminated area with plants specially selected for a high potential to uptake the targeted contaminant. Another remediation strategy is to surround an underground zone of contaminated soil by specially selected plants, to prevent the further distribution of contaminants because of the hydrostatic barrier formed by roots [152].

In spite of significant progress in the creation of phytoremediation applications, there are still significant gaps in the evaluation of potential plants for phytoremediation. For instance, there is lack of knowledge on the role and function of leaves in atmospheric cleanup. Published data and the data of the present authors accumulated over the last 30 years directly indicate that leaves absorb pollutants [515] and herbicides dissolved in water [58] but existing knowledge needs to be supplemented with new data.

Phytoremediation on the basis of recent understanding includes the following technologies: phytoextraction, rhizofiltration, rhizodegradation, phytodegradation, phytotransformation, phytostabilization, etc. In Table 4.4 the levels and aims of the application of phytoremediation technologies are presented.

Table 4.4. Phytoremediation overview [151]

Mechanism	Process Goal	Media	Contaminants	Plants	Status
1	2	3	4	5	6
Phytoextraction	Contaminant extraction and capture	Soil, sediment, sludges	Metals: Ag, Cd, Co, Cr, Cu, Hg, Mn, Mo, Ni, Pb, Zn; Radionuclides: ^{90}Sr, ^{137}Cs, ^{239}Pu, ^{238}U, ^{234}U	Indian mustard, pennycress, alyssum, sunflowers, hybrid poplars, *thlaspi*	Laboratory, pilot, and field applications
Rhizofiltration	Contaminant extraction and capture	Groundwater, surface water	Metals, radionuclides	Sunflowers, Indian mustard, water hyacinth	Laboratory and pilot-scale
Phytostabilization	Contaminant containment	Soil, sediment, sludges	As, Cd, Cr, Cu, Hs, Pb, Zn	Indian mustard, hybrid poplars, grasses	Field application
Rhizodegradation	Contaminant destruction	Soil, sediment, sludges, groundwater	Organic compounds (TPH, PAHs, pesticides, chlorinated solvents, PCBs)	Red mulberry, grasses, hybrid poplars, cattail, rice	Field application
Phytodegradation	Contaminant destruction	Soil, sediment, sludges, groundwater, surface water	Organic compounds, chlorinated solvents, phenols, herbicides, munitions	Algae, stonewort, hybrid poplars, black willow, bald cypress	Field demonstration

Table 4.4. (cont.)

1	2	3	4	5	6
Phytovolatilization	Contaminant extraction from media and release to air	Groundwater, soil, sediment, sludges	Chlorinated solvents, some inorganics (Se, Hg, and As)	Poplars, alfalfa, black locust	Laboratory and field application
Hydraulic control (plume control)	Contaminant degradation or containment	Groundwater, surface water	Water-soluble organics and inorganics	Hybrid poplars, cottonwood, willow	Field demonstration
Vegetative cover (evapotranspiration cover)	Contaminant containment, erosion control	Soil, sludge, sediments	Organic and inorganic compounds	Poplars, grasses	Field application
Riparian corridors (non-point source control)	Contaminant destruction	Surface water, groundwater	Water-soluble organics and inorganics	Poplars	Field application

4.2.1 Phytotransformation

Phytotransformation (or phytodegradation) is a completely natural technology. These terms refer to the uptake, accumulation, and transformation of organic toxic contaminants from soil, air, and water reservoirs by different plants. Plants take up the contaminants and, depending on their structure and hydrophilicity, either partially transform them to more reactive intermediates or directly conjugate them in plant tissue. In other words, as a result of uptake, plants store the contaminants in cellular structures or degrade them to regular metabolites or finally to CO_2 (deep oxidation) and water.

The application of phytotransformation is especially effective for the remediation of sites polluted by the following contaminants: petroleum wastes, explosive wastes, chlorinated solvents, pesticides, etc. Plants easily take up toxic organic compounds of different structures from soil or water or degrade them by enzymes. To repeat: uptake greatly depends on hydrophobicity, solubility, and polarity of the contaminants. According to some calculations [44], uptake of contaminants by plants from contaminated soils at a shallow depth is an effective process for moderately hydrophobic organic chemicals (octanol-water partition coefficient, log $K_{ow}=1$ to 3.5). Such contaminants are easily taken up and translocated by several dozen trees species [367]. Compounds in this category include chlorinated solvents and short-chain aliphatic chemicals. Chemicals with a hydrophobic nature (log $K_{ow} > 3.5$) are strongly bound to the plant roots, and their translocation within the plant is a complicated process [97, 438]. The efficiency of contaminant movement in plants depends on several factors such as transpiration rate, concentration of contaminant, type of plant, etc. As a rule, more than 80% of the xenobiotics penetrating into a plant cell undergo a conjugation process. It is important to note that conjugated compounds are almost always less toxic than the original unconjugated contaminants. Nevertheless, in spite of some detoxification effect, this pathway cannot be considered as a real detoxification process (due to the preservation of the complete structure of the toxic compound) but rather as a temporary process to evade the toxicity of the xenobiotic.

The uptake of any particular contaminant greatly depends on the plant species, season, type of soil, water, chemical nature of the contaminant, contaminant concentration in the soil, and some other physical, chemical, and agronomical factors. Conclusions concerning any contaminant uptake by a particular plant species cannot be applied to other plants, even to those belonging to the same genus. According to some data [391] 21% of PCP in the soil was found in the roots of grasses and 15% in the shoots after 155 days of cultivation. In another study [27] several plants showed a low PCP uptake ability. Concerning the metabolic potential of plants, it should be emphasized that plants are able to absorb hydrocarbon contaminants of different structures – PAHs, PCBs, pesticides, solvents, etc. [220, 278, 511] – via both their root system and leaves and detoxify them by depositing their conjugates created by transferases in internal cell structures and partially metabolizing them by promoting deep oxidation by oxidative enzymes. The primary process is the most important one for the preservation of the environment and consists in deep oxidation of organic contaminants to regular metabolites of the plant cell or to carbon dioxide and water. The cytochrome P450-containing monooxygenase system, peroxidase, and phenoloxidase, acting step by step, are the main enzymes that oxidize organic toxic compounds. In the literature, there is other evidence indicating the degradation of different hydrocarbons to CO_2, or regular cell metabolites by plants [220, 278]. Newman and coauthors [354] have reported the mineralization of such difficult-to-degrade contaminants as TCE, by poplar.

4.2.2 Phytoextraction

Phytoextraction deals with the absorption of heavy metals by roots and their translocation within the plants. Exudates rich in organics control the pH in the microenvironment of the root system, creating appropriate conditions for the prolonged absorption of metals by roots, following metal penetration into root cells.

Phytoextraction is also used for the treatment of polluted sediments and sludge, and to a lesser extent for the cleaning of water. Translocation of metals to the plant parts above ground allows elaboration of cost-effective technologies for their extraction and elimination from the soil. In the literature, examples have been reported indicating the effective use of phytoextraction technologies in the remediation of soils contaminated by cadmium, nickel, zinc, arsenic, selenium, and copper. This technology is moderately effective with cobalt, manganese, and iron, but until now less demonstrably effective for lead, chromium, and uranium [151, 397, 422, 423, 437]. The efficiency of phytoextraction is determined by two factors:
- The accumulation factor, indicating the ratio of metal concentrations in the plant organs (shoots, roots) and in the soil.
- The seasonally harvestable plant biomass.

These two factors are important for the design of phytoextraction technologies. Perennial, fast growing plants, with a well-developed root system and a luxuriant biomass above ground, are most advantageously used in phytoextraction. These are mainly leafy plants growing well under warm and cold conditions. However, plants with low biomass but high metal accumulation abilities, such as the majority of algae and marsh plants could also be successfully used in phytoextraction (Table 4.5).

Table 4.5. Metal-accumulating plants used in phytoextraction technologies

Plants	Heavy metals	References
Indian mustard (*Brassica juncea*)	Pb, Cr(VI), Cd, Cu, Ni, Zn, Sr, B, Se	[397, 422, 437, 455]
Alpine pennycress (*Thlaspi caerulescens*)	Ni, Zn	[48]
Pennycress (*Thlaspi rotundi-folium* ssp. *cepaeifolium*)	Pb	[350]
Alyssum (*Alyssum wulfenianum*)	Ni	[399]
Canola (*Brassica napus*) Kenaf (*Hibiscus cannabinus* L. cv. Indian) Tall fescue (*Festuca arundinacea* Schreb cv. Alta)	Se	[22]
Poplar (*Populus* sp.)	As, Cd	[383]
Sunflower (*Helianthus annuus*)	Cs, Sr	[3]
Sudangrass (*Sorghum vulgare* L.) Alfalfa (*Medicago sativa*) Maize (*Zea mays*)	Pb, Zn, Hg, Ni	[151]

The disposal of plants and associated biomass used in phytoextraction is of great importance. There exists a real danger of introducing contaminants into the food chain via the consumption of contaminant-containing plant biomass by livestock or wildlife. Hence it is an obvious necessity to monitor and test plant biomass for the presence of toxic compounds. The harvest of contaminated plant biomass and its disposal as hazardous waste requires special regulations (cf. Resource Conservation and Recovery Act (RCRA) US). To evaluate completely the effectiveness and depth of contaminant transformations, residues of contaminants and the intermediates of their transformations, which sometimes exceed the toxicity of the initial chemicals, should be analyzed appropriately. In evaluating the effectiveness of a phytoremediation application, such monitoring must be performed on all plant biomass, as well as for soil and groundwater, and only when favorable results of all these determinations have been obtained it can be concluded that the clean-up goals and objectives have been reached.

4.2.3 Rhizofiltration

This ecotechnology emphasizes the removal of metals derived from aquatic environments such as damp soil and groundwater by the rhizosphere. Two strategies are available for its effective exploitation. The first one consists in sorption of metals by the root system and their removal – depending on physiological and biochemical characteristics of the plant. This technology is especially effective due to the ability of some plants to absorb large quantities of metals (lead, chromium, etc.) from soil water, accumulated or passing through the root system zone. Another strategy is the construction of wetlands for the treatment of contaminated water. Wetlands have been effectively used in treating metals and some organic contaminants [542].

Rhizofiltration is effective for the treatment of large volumes of wet soil and water containing low concentrations of contaminants. In most cases plant species specially selected for this aim are grown hydroponically and transplanted to a floating system in which the plant roots are in intimate contact with the contaminants. Rhizofiltration was used to deal with uranium wastes at Ashtabula, Ohio, and uranium-contaminated water in a pond near the Chernobyl nuclear plant in Ukraine; in both cases sunflowers were planted [437].

4.2.4 Rhizodegradation

Rhizodegradation takes place in the rhizosphere, by creating beneficial conditions for microbial growth and development in the rhizosphere. This process is greatly supported by exudates containing a wide spectrum of organic compounds. Rhizodegradation implies the phytostimulation or enhanced rhizosphere remediation by the root system.

Activation of the rhizodegradation processes is connected with the increase in soil active components, such as consortia of soil microorganisms, mycorrhizal fungi biomass, and a spectrum of exuded organic compounds (small molecular mass organics, enzymes, carbohydrates, etc.). Some trees, e.g., orange, apple and mulberry, excrete flavonoids and coumaric acid into the soil, stimulating the degradation of polychlorinated biphenyls [120, 191]. The degradation of herbicides is stimulated by kochia [374], knotweed, crabgrass [8], pine [187] etc. The population of nitrogen-fixing microorganisms in polluted soil increases appreciably after the additional introduction of *Azotobacter* species and free-living nitrogen-fixing actinomycetes [292]. These organisms indirectly participate in the remediation, especially longterm, providing plant and soil microorganisms with nitrogen-containing compounds. The participation of fungi in the symbiotic association is also very important, since due to their unique enzyme systems they degrade soil organics that cannot be transformed by prokaryotes. The partial transformation of organics by soil microbial consortia that thereby decrease the penetration barrier of the roots is also of great importance.

4.2.5 Hydraulic control

Hydraulic control (phytohydraulics) refers to groundwater transport driven by plant transpiration [151]. It involves of plants with high respiration rates that take up large amounts of water and contaminants, thereby preventing their further migration. The technology is effective only if the selected plants have a well-developed root system. The most convenient plants for this application are poplar, birch, willow, eucalyptus, etc. The plant action is determined by temperature, and, thus, the activity of these plants is significantly decreased in winter.

4.2.6 Phytostabilization

Phytostabilization is based on plants' ability to immobilize metals in the soil by sorption, precipitation and complexation. The process stabilizes the soil matrix, and minimizes soil erosion and migration of sediments [152]. Phytostabilization is especially applicable to contamination for which on-site fixation of metals is needed. For the realization of phytostabilization technologies, vigorously growing plants performing hydraulic control and immobilization are the most convenient.

4.2.7 Phytovolatilization

Phytovolatilization refers to the uptake and subsequent unchanged release of VOCs into the atmosphere, as in the case of TCE, or their transformation via transpiration [264]. This technology does not eliminate volatile contaminants from the environment but removes them from the soil and groundwater.

Once created, phytovolatilization and other phytoremediation plantations should be monitored and their effectiveness (i.e., their activity) measured periodically.

To ensure healthy growth and normal development of plants, regular agrotechnical operations should be carried out. Weed control and irrigation are especially important. As a result of the proliferation of particular weeds, predators, and diseases, a significant reduction in the activity of phytoremediating plants can occur. Weed infestation could be avoided by using herbicides, which will, however, necessitate further planting to then eliminate the herbicide residues. Irrigation water must compensate water loss by evaporation and respiration. Wetland systems are a very useful asset of plantations for water distribution and the removal of surplus vegetation and contaminants.

4.2.8 Cleaning of the air

In most phytoremediation technologies a great deal of attention is paid to the cleaning up of soil and water, yet the search for means and technologies for the cleaning of contaminated air is not less important. One recalls that in the past, entire cities have been abandoned because of bad air. Paestum is a good example. The principle strategy in air cleanup is the creation of technologies that reduce the net emission of harmful gases from industry and vehicle exhausts into the atmosphere to a minimum. The problem has been strongly politicized: governments of many countries have signed and ratified agreements, declarations and conventions pledging to restrict and strongly control toxic emissions into the atmosphere (the Montreal protocol, 1987; the Kyoto protocol, 1992; the Rio declaration, 1992; the Orhus declaration, 1997; etc.).

Air contamination motivates the wide use of different technological approaches to diminish the problem, including the exploitation of plants as a natural tool.

At the Durmishidze Institute of Biochemistry and Biotechnology of the Georgian Academy of Sciences, since 1960 investigations have been conducted to characterize the potential of plants to absorb and metabolize xenobiotics. This work enabled the evaluation of the potential of over 100 species of annual and perennial plants (herbaceous and shrubby trees) to absorb via their leaves and metabolically degrade over 150 different contaminants from the following classes: alkanes, alkenes, aromatic and polycyclic hydrocarbons, their nitro- and chloro-derivatives, alcohols, phenols, aldehydes, ketones, etc.). Based on these experiments and as a result of over twenty years of collaboration with other laboratories, the concept of the "Green filter" has been elaborated, envisaging the purposeful planting of selected plants around the sources of contamination, i.e. along highways, chemical and metallurgical factories, etc. to reduce the stream of environmental contaminants [510].

Nowadays, many physical and chemical methods and technologies of contaminated air purification exist [102]. The biological principles of atmospheric cleaning have also been successfully realized in specially constructed bioreactors: biofilters, bioscrubbers, etc. Biofilters contain microorganisms (peat or synthetic material), immobilized on a solid base, capable of degrading contaminants absorbed by filtering material from the air. The principle of action of bioscrubbers is as follows: contaminants are adsorbed by a water phase, which moves through the bioreactor containing active silt or microorganisms immobilized on a solid carrier. Related technological approaches using microorganisms include biofilters with irrigated layers [287].

In contrast to microorganisms, plants do not require specially constructed apparatus. Over 40% of land is still covered by plants, thus providing sufficient contact of plants with gaseous air contaminants. While dealing with plants, the most important action is the appropriate selection of plant species taking up and transforming gaseous contaminants. The assembly of plant communities specially selected to create a "Green filter" corresponding to the source of chemicals around the source of the contamination follows criteria established according to the peculiarities of the location, soil type, direction and speed of the wind, type and degree of the contamination, and the physical and chemical characteristics of the contaminants. Among several possible types of "Green filter", the most generally recommended has the structure of three plant layers: the lowest (first) layer consists of herbaceous annual plants; the second of shrubby plants; and the third of tall woody trees. The width of each plant line depends on the intensity of the stream of contaminants and their nature. This particular "Green filter" structure is conditioned by the fact that atmospheric contaminants, which tend to be heavier than air, are concentrated close to the soil surface (in a layer about 1.5 m above the ground) and have the most intensive contact with herbaceous plants. The second plant line also intensively contacts the polluted air and significantly decreases the concentration of contaminants. The third plant line (trees) mechanically protects the "Green filter" zone from excessive wind [510] and other undesirable influences in addition to absorbing environmental contaminants.

Plants can very effectively hinder the emission of environmental contaminants in the exhaust gases of motorcars: carbon monoxide, nitrogen and sulfur oxides, aromatic hydrocarbons, etc. A "Green filter" 30 m wide and consisting of five layers (four lines of shrubs, mulberry and fustic, 70 cm high, and a fifth line of maple, birch, and elm 6–8 m high, plant age 12 years) planted along a highway reduces carbon monoxide concentrations in the air by 60–70%, in comparison with the concentration of these aerial contaminants elsewhere along the highway [510].

Table 4.6. Groups of higher plants ranked according to their ability to assimilate aromatic hydrocarbons (benzene and toluene) via their leaves [510]

Group of plants	Amount of absorbed aromatic hydrocarbon by 1 kg of fresh leaves, for 24 hours, mg	Plants	
1	**2**	**3**	**4**
Strong absorbers	1.0 – 10.0	Maple (*Acer campestre*) Apple (*Malus domestica*) Oleaster (*Elaeagnus angustifolia*) Zelkova (*Zelkova caprinifolia*) Locust (*Robina pseudoacacia*) Poplar (*Populus canadensis*) Caucasian pear (*Pyrus caucasica*) Ryegrass (*Lolium perenne*) Walnut (*Juglans regia*) Lilac (*Syringa vulgaris*)	Almond (*Prunus amygdalus*) Weeping willow (*Salix babylonica*) Cherry (*Cerasus avium*) Catalpa (*Catalpa bignonioides*) Amorpha (*Amorpha fructicosa*) Oriental plane (*Platanus orientalis*) Cherry (*Cerasus vulgaris*) Sophora (*Sophora japonica*) Chestnut (*Castanea sativa*)
Average absorbers	0.1 – 1.0	Alder (*Alnus barbata*) Asp (*Populus tremula*) Elm (*Ulmus filiacea*) Ash (*Fraxinus excelsior*) Tea (*Thea sinensis*) Persimmon (*Diospyros lotus*) Bay laurel (*Laurus nobilis*)	Gleditschia (*Gleditischia triacanthos*) Kidney bean (*Phaseolus vulgaris*) Pine (*Pinus*) Pine (*Pinus eldarica*) Thuja (*Tuja*) Apricot (*Prunus armeniaca*) Vine (*Vitis vinifera*)

Table **4.6.** (continued)

1	2	3	4
Weak absorbers	0.001 – 0.1	Norway spruce (*Picea abies*) Mulberry (*Morus alba*) Lime-tree (*Tilia caucasica*) Reed (*Phragmites communis*) Maize (*Zea mays*) Wild plum (*Prunus divaricata*) Kiwi (*Apteryx australis*) Aloe (*Aloe*) Medlar (*Mespilus germanica*) Rose (*Rosa*) Cypress (*Cupressus sempervirens* var.Pyramidalis)	Geranium (*Pelargonium roseum*) Privet (*Ligustrum vulgare*) Fig (*Ficus carica*) Pomegranate (*Punica granatum*) Rhododendron (*Rhododendron ponticum*) Peach (*Persica vulgaris*) Potato (*Solanum tuberosum*) Tomato (*Lycopersicum esculentum*) Pussy-willow (*Salix alba*) Cherry-plum (*Prunus vachuschtii*)

Adsorption and uptake by plants of aerosols, dust, and smoke microparticles, containing both inorganic and organic contaminants such as aromatic hydrocarbons (benzene and toluene), emitted in huge amounts into the atmosphere as exhausts and industrial smokes, have a great ecological impact. The total area of plant leaf cover, stomata density, intensity of carbon dioxide uptake and other less important factors determine the uptake intensity of deleterious atmospheric gases. It is very important to realize that the ability to take up and degrade toxic compounds significantly differs among plants. Quantitatively the difference between similarly sized plants might range from 1 microgram to 10 milligrams of aromatic compounds per 24 hours (Table 4.6) and hence the difference between plants' detoxification abilities for particular contaminants may easily exceed four orders of magnitude. According to the data of Table 4.6 plants can be divided into weak, average, and strong absorbers of benzene and toluene [510]. Such differences in contaminant assimilation potential create a good basis for the rational selection of the most appropriate plant species for use in the different soil-climatic areas of our planet and make the universal use of phytoremediation technologies possible.

4.3 Transgenic plants in phytoremediation

During the last decade phytoremediation has evolved from a conceptual methodological approach into an ecologically important commercial technology for environmental cleanup from organic and inorganic contaminants. For the successful realization of phytoremediation the synergetic action of microorganisms and plants is the basis. For an effective practical implementation of these technologies the most important factor is the presence of suitable plants capable of actively assimilating contaminants. The efficiency of the phytoremediation processes is determined mostly by the plant's ability to assimilate or accumulate (or both) organic and inorganic contaminants in cellular structures and to accomplish deep oxidative degradation of organic xenobiotics. Progress in the development of phytoremediation technologies applied to environments contaminated with organic contaminants is significantly ahead of developments in the assimilation of inorganic contaminants and radionuclides, simply because work began earlier on selecting suitable plants with desirable features such as adaptation to the particular soil-climatic zone productivity (fast growing, large biomass formation), presence of the corresponding physiological (transpiration ability) and morphological (development of an appropriate leaf and root system), characteristics adaptation to field conditions, presence of the needed broad enzymatic systems, etc.

These specificity characteristics, and possibly others, are the basis of assimilation and deep degradation of organic contaminants by plants. In order to increase the ecological potential of plants, definite progress has already been achieved in the cloning of genes participating in contaminant transformation. According to the literature a number of modified plants having especially high contaminant accumulation ability and a correspondingly large intracellular volume to deposit metabolite–xenobiotic conjugates have been created. Investigations in this direction are presently intensively carried out in more then 50 laboratories in various countries. Since it is impossible to cover all ecologically important genetically modified plants within this chapter, some examples of the existing approaches are presented.

For about two decades the experimental data concerning the ability of plants to clean soils and groundwaters from inorganic contaminants have been intensively discussed in the literature. Phytoextraction of heavy metals under typical site conditions is the cheapest and safest method, and has drawn the attention of scientists, practicing agrarians and ecologists. The progress achieved in growing special forms of plants, distinguished by their high efficiency for the phytoextraction of heavy metals, will be briefly described below. According to the literature, [208, 320, 321] transgenic plants are often characterized by abnormally high phytoextraction abilities. Since well-known genetic engineering techniques have been used to create plant transformants, the detailed methodological approaches will not be described here. Attention will be paid to the use and characterization of important detoxification abilities of the plant transformants.

Plant genetic engineering investigations directed towards raising the efficiency of phytoremediation have been carried out especially intensively for the last decade. The evaluation of specific features of the modified plants was mostly performed in greenhouses, i.e., within small, controlled zones for ecological reasons. The first field experiments were initiated at the beginning of the new millenium in the USA. Recombinant plant creation (currently more than 100 have been reported) proceeds in several different directions.

Among the large variety of plants with perspectives for phytoremediation, the poplar family attracts special interest. Because of its strong root system it is characterized by high absorption ability. As a result of multiple gene-engineered modifications of this plant various transformants have been generated, some of them yielding convincing evidence for effective practical use. One deals with enriching the poplar genome with a bacterial gene encoding γ-glutamylcysteine synthetase, the synthesis of which is the rate-limiting factor in glutathione biosynthesis [205]. For comparison a wild hybrid of poplar (*Populus tremula* x *alba*) together with two transgenic clones characterized by γ-glutamylcysteine supersynthesis were used. The plants were cultivated in soil containing the chloracetinyl herbicides metolachlor and acetochlor. The growth of all plants was suppressed to different degrees under the influence of the herbicides, but the transgenic plants, distinguished by γ-glutamylcysteine supersynthesis, accumulated biomass much more intensively than the wild poplar hybrids. The transgenic clones were also distinguished by increased activities of γ-glutamylcysteine synthetase and glutathione-S-transferase (GST) in the leaves, indicating the close relationship of these biosynthetic processes.

GST is a widely distributed enzyme in plants, participating in normal metabolic processes as well as protecting plants from stress factors. In growing transgenic plants for phytoremediation purposes, the gene encoding this enzyme is frequently the object of attention. Genes of three recombinant GSTs with molecular masses of 26, 27, and 29 kD from the genome of *E. coli* have been cloned into maize. The monomeric forms were aggregated into dimers in order to be active; from these three monomers four dimeric forms of GST were found to participate in the conjugation of herbicides and different xenobiotics [302]. Among the several known molecular forms of GST the greatest attention has been reserved for GSTs II–II and III–III. GST II–II protects protoporphyrin from degradation, and the enzyme form III–III protects from autooxidation. Besides, it has been shown that definite substrate specificity is characteristic for each molecular form of GST. It could be that such specificity on the level of enzyme activity is the important rate-limiting factor for the detoxification process.

Transgenic plants have also been studied in connection with particular soil-climatic conditions and contaminants. For this purpose, the widely distributed explosive TNT has generally been chosen. In order to increase the degradability of TNT and similar compounds, several transgenic tobacco plants, containing the gene of the bacterial enzyme pentaerytrole tetranitrate reductase (EC 1.6.99.7) have been created [179]. This transgenic tobacco has been analyzed for its ability to assimilate the residues of explosive compounds such as TNT and trinitroglycerine. Seedlings of the transgenic plants extracted explosives from liquid phases much faster, accomplishing denitration, than the seedlings of common forms of the same plants, in which growth was inhibited by the contaminants [213]. Transgenic tobacco differed substantially from the common plant by its tolerance and fast uptake and assimilation of significant amounts of TNT. Analogous experimental results have been obtained with other plant species [214, 285].

There are various publications [123, 205, 275, 302] on the improvement of plant detoxification abilities by the cloning of genes encoding transferases and oxidases, which intensively participate in contaminant transformation processes. There is no doubt that the genetically modified plants are often significantly more promising detoxifiers than their original unmodified precursors. If we were to try to imagine the ideal plant for phytoremedial applications, evidently the picture would be the following: as a basic requirement, such a plant should have a well-developed root system and a strong transpiration stream; it must intensively form biomass, with a high resistance towards organic and inorganic toxic compounds; it must form molecular conjugates and have the required potential for their accumulation into cellular organelles, and it must be able to degrade organic contaminants irrespective of their structure. In addition, it must secrete exudate amounts sufficient to enhance microbial degradation in the rhizosphere.

Characteristics of plants suitable to phytoremediate inorganic contaminants present another picture. Scanty information on the molecular mechanisms of plant tolerance to heavy metals, and of data pointing to the genes responsible for their uptake, complicate the rational modification of plants to increase their detoxification potential. A number of original and review works published recently [229, 309, 319, 472] are devoted to the discussion of these and other problems concerning the uptake of inorganic contaminants. Some of these publications describe transgenic plants characterized by enhanced tolerance to cadmium and lead (70–75 mM), which inevitably point to their hyperaccumulation potential. Modified plants with double the lead content of the original ones have been described [380].

A review of publications on genetic engineering [179, 205, 214, 302, 363] shows that in a number of cases an increase in the detoxification potential of transgenic plants, i.e. their ability to assimilate organic contaminants and absorb heavy metals, has been observed. Investigations in this direction will doubtless be continued in the near future, and the results will be more substantial when considered from the viewpoint of eventual practical use. The benefit of these investigations would be more impressive if all aspects of the quite complicated and multistage detoxification process were better understood with regard to plant physiology and biochemistry. This would allow the creation of a more rational and effective strategy for the cloning of genes, especially in the case of inorganic contaminants.

It is noteworthy and regrettable that the majority of genetic engineering studies are connected with GST and other enzymes participating in the conjugation process, and very little attention is paid to enzymes catalyzing the immediate degradation of organic contaminants. This indicates that plants are still widely considered as organisms merely accumulating contaminants as conjugates with cellular compounds. Recent investigations simply point to the rather high activity of plants directed to the degradation of various organic substances, including aromatic xenobiotics. Few investigations in genetic engineering are devoted to peroxidase, one of the general enzymes participating in the oxidative degradation of xenobiotics. The participation of phenoloxidase has scarcely been considered in detoxification processes, yet this enzyme plays a very important role in the oxidative degradation of xenobiotics, especially those with an aromatic structure. Dozens of publications have been devoted to the transfer of genes of monooxygenases, the enzyme system carrying out the initial oxidation of organic contaminants (for instance, introduction of hydroxyl groups) [123, 275]. Great interest is connected with the transfer of cytochrome-P450 genes to different plants [363]. Transgenic plants containing an extra cytochrome-P450 gene are generally characterized by increased resistance to herbicides of a diversity of structures and have clearly observable improved detoxification potentials.

In spite of some problems with public acceptance of and legislation concerning transgenic plants, we should consider these organisms as a very important biological tool having tremendous ecological potential.

A comprehensive study of plant detoxification mechanisms at the molecular level would greatly improve our understanding of the whole process of phytoremediation and allow the creation of more efficient plant transformants, which among other advantages would be capable of effective interaction with the microbial community in the rhizosphere and synergism in contaminant degradation.

4.4 The cost of phytoremediation technologies

Phytoremediation is a completely natural process based on the joint detoxification action of plants and microorganisms. Phytoremediation technologies are economically competitive, compared with existing conventional ones. Dozens of scaled-up examples have demonstrated the superiority of plant-based remediation technologies, mainly due to the following reasons [152]: phytoremediation, being a natural, solar energy-driven process, does not require any additional energy or significant material or other input; phytoremediation takes place *in situ* and requires no digging or hauling; little mechanical equipment is needed to operate phytoremediation. The cost components for the implementation of phytoremediation technologies include:

- Detailed characterization of the soil type, water content, type of contaminant(s), concentration of contaminant(s) in the soil, etc. of the polluted site
- Selection of appropriate plants and microorganisms
- An irrigation system
- Capital cost, materials, monitoring, including required instrumentation, indirect costs, etc.
- Operation and maintenance (labor, materials, chemicals, laboratory analyses, etc).

Phytoremediation offers cost advantages, but it should be underlined that the time needed for full remediation is typically lengthy. Tables 4.7 and 4.8 give estimates of the costs of phytoremediation as compared with conventional technologies.

Table 4.7. Estimated cost savings through the use of phytoremediation rather than conventional treatment, according to EPA data [152]

Contaminant and matrix	Phytoremediation		Conventional treatment		Projected savings
	Application	Estimated cost	Application	Estimated cost	
Lead in soil (1 acre)	Extraction, harvest, and disposal	$ 150,000–250,000	Excavate and landfill	$ 500,000	50–65%
Solvents in groundwater (2.5 acres)	Degradation and hydraulic control	$ 200,000 for installation and initial maintenance	Pump and treat	$ 700,000 annual operating cost	50% cost saving by 3rd year
Total petroleum hydrocarbons in soil (1 acre)	*In situ* degradation	$ 50,000–100,000	Excavate and landfill or incinerate	$ 500,000	80%

Table 4.8. Cost advantage of phytoremediation (rhizosphere bioremediation) of soils using fine-rooted grasses, compared to other techniques [437]

Type of treatment	Range of costs, $/ton
Phytoremediation	10–35
In situ bioremediation	50–150
Soil venting	20–220
Indirect thermal treatment	120–300
Soil washing	80–200
Solidification/stabilization	240–340
Solvent extraction	360–440
Incineration	200–1500

4.5 Phytoremediation – an effective natural tool for a healthy planet

Plants have always had a decisive importance for the existence of mankind. By providing renewable resources of food, fuel and fiber during millennia plants have been supporting the existence of human life and the advancement of primitive societies. During that time, mankind, always governed by climatic and other environmental conditions, was able to apply and select from the great variety of plants on earth at first intuitively and later scientifically. At that time plants were already considered as absolutely and components of the environment and for the existence of mankind. Much later, in parallel with development of such natural sciences as botany, plant physiology, plant biochemistry etc., the participation of plants in such vitally important processes as photosynthesis and fixation of N_2 was demonstrated. The use of plant secondary metabolites, whose efficacy was established by a laborious process of trial and error (Darwinian selection applied to discovery), for traditional medicinal reasons has apparently been important since the dawn of human civilization. Nevertheless, the traditional strong belief in the functional importance of plants became gradually and metaphorically etiolated. As a result of urbanization, industrialization, military actions, and other anthropogenic factors global ecosystems have deteriorated and in many regions of the planet have reached a critical level, below which their viability is imperilled or has already been lost forever due to the ignorant and greedy management of natural resources. A number of chemical, physical, and mechanical technologies and other know-how enabled some local environmental problems to be solved, but had no appreciable influence on the planet's ecology as a whole. As was later clarified, much more effective is the power, represented in nature by both lower and higher plants, directed towards the degradation of organic contaminants of diverse natures. Nowadays, the activities of microorganisms directed to the degradation of contaminants of diverse structures are widely used in practice (wastewater treatment, elimination of contaminants from groundwater, purification of sewage, decomposition of organic contaminants in soil, etc.). In these technologies actinomycetes, bacteria, yeasts and mycelial fungi assimilate a broad spectrum of organic contaminants, due to their extensive genetic resources leading to significant or even full detoxification. Some of these technologies, notably for wastewater treatment, were in fact already well established a century ago in advanced and innovative economies such as in Great Britain.

Higher plants for a long time had not been considered as organisms participating in detoxification processes: plants were thought of as organisms with a naturally limited potential for contaminant conjugation and accumulation. Investigations carried out mostly during the last decade have revealed the potential of plants to absorb and decompose organic contaminants and accumulate inorganic contaminants from soil, water, and air. Depending on the nature of the organic xenobiotic and the plant species, typically 1 kg of green biomass takes up from the air daily amounts ranging from micrograms to dozens of milligrams of contaminants [510]. Plants possess universal cleaning up abilities for purifying soil, groundwater, and air. Higher plants can be considered as ecological tools for eliminating heavy metals from the soil by translocating them from below to above ground. Some plant species, and especially their transgenic modified forms, are known as hyperaccumulators of metals, as for instance, Indian mustard, poplar, tobacco, *Thlaspi*, *Arabidopsis*, etc. [183, 208, 320, 321, 380].

In order to expand the frontiers of phytoremediation technologies great attention is now being paid to the use of transgenic plants with significantly increased detoxification potentials compared with the parent (unmodified) plants. The improvement of remediation abilities is directly connected with increased accumulation potential for toxic heavy metals (As, Pb, Hg, Cd, etc.), and the degradation of particular contaminants that are barely normally degradable, as for instance POPs. It might be that such plants represent the future for the cleaning up of the environment. Limitations in the large-scale application of transgenic plants arise through ecological reasons, i.e., the threat to the biodiversity of local (endemic) plants. The technological goal to avoid this problem is to achieve complete male and female sterility of the genetically modified plants.

Phytoremediation is a multicomponent process, combining the use of plants and microorganisms in landscape management. It implies different approaches based on multiple factors beginning with plant and microbial physiology and ending with genetic engineering.

As has been shown in practice, phytoremediation, as a long-term technology, works best with shallow contamination of the soil. Elimination of contaminants located deeper than two meters is connected with limitations in time, since mass transfer processes at that depth proceed much more slowly than in the upper parts of the soil. Hence extraction by roots and the subsequent transport may become the rate-limiting factor of the whole process. Therefore, plant-microbial action-based technologies would need excessive time to achieve a satisfactorily clean standard of soil. According to some existing views [161], phytoremediation as a final "polishing step" must follow other technologies such as excavation and treatment or disposal.

Another situation in which phytoremediation is not yet successfully applied is to treat a high concentration of soil contaminants such as PCBs and dioxins, which completely prevents the growth of plants. In such extraordinary cases phytoremediation alone cannot clean the soil up in any realistic time frame. But stating categorically that the application of phytoremediation technologies cannot work in these cases may be a serious mistake. First of all, plants significantly differ in their detoxification abilities and we should not ignore the possibilities of physiological adaptation in plants, possibly vastly accelerated by hitherto undeveloped technology, leading to the selection of plants (annual, perennial) with an increased resistance to a single contaminant or a group of contaminants of similar structure. Prior selection of such plants could have unexpectedly positive results. Secondly, especially selected soil microorganisms of different taxonomic groups could be set to attack contaminants, resulting in a decrease of their concentrations to the levels allowing plants (of a certain species) to start to grow at polluted sites. Finally, the use of transgenic plants and possible genetically modified microorganisms may also have a great influence on the possibilities of remediation of sites with an especially high concentration of persistent pollutant.

Plants are very promising detoxifiers allowing the creation of safe technologies around or along hotbeds of contamination (Green filter, vegetation cap, phytoremediation cover, hydrologic control, evapotranspiration cover or any other plant–based technology) — ecologically friendly, definitely positive, and of significant local importance. It should be clearly realized that nowadays technologies for enabling the maximum potential of plants in combination with microorganisms for accumulating and degrading organic contaminants of different structures are definitely within reach, and allow us to widen our understanding of the phytoremediation potential and realize its use on a scale significantly exceeding any existing local or even national level. Since there is no universal ecotechnology to clean up all kinds of contamination, phytoremediation solely or in combination with other convenient types of technologies should be considered as the best solution of the problem. Elaboration of a new ecological concept, unifying experience accumulated for the last three decades in particular and the realization of new plant-based approaches on a world scale should lead to a beneficial increase of the ecological potential of the whole planet.

Creation of a good foundation for the realization of a new, ecological concept, namely "Plants for a healthy planet", requires additional activities necessary for the world-scale realization of the concept. To realize such a project the following important aims should be considered:

- **Investigations needed for the detailed biochemical and physiological analysis of the whole process of phytoremediation – a group of innovative technological approaches (detailed characterization of all types of oxidases catalyzing degradation of organic contaminants, and of transferases and other enzymes participating in the detoxification process etc.) should be continued; the creation of new, modified, genetically stable, environmentally safe, highly effective vegetation (growing under different climatic conditions) trees, grasses, legumes, terrestrial or aquatic plants; the selection of microorganisms (bacteria, fungi, actinomycetes) with special attention to symbiotic action and scaling up of phytoremediation processes.**
- **International recognition and worldwide support for the realization of a new ecological concept (international cooperation, including financing; involvement of such international organizations as NATO, UNESCO, the U.S. Environmental Protection Agency (EPA), the European Environmental Agency (EEA), the United Nations Environment Program (UNEP), and national and international green movement (Greenpeace, Global Greens, etc.)**
- **Organization of scientific branches in all continents and including remote districts, responsible for the establishment of a world pollution and remediation survey and the creation of corresponding mathematical models of existing hotspots, contaminated sites, and distribution of contaminants; and the creation of approaches for their remediation**
- **Introduction of a special interdisciplinary education program for phytoremediation at universities and institutes of technology.**

References

1. Adamia G, Khatisashvili G, Varazashvili T, Pruidze M, Ananiashvili T, Gvakharia V, Adamia T, Gordeziani M (2003) Determination of the type and rate of soil contamination with heavy metals and organic toxicants on the territories of military proving grounds in Georgia. Bull Georg Acad Sci 167: 155–158

2. Adamia G, Ghoghoberidze M, Graves D, Khatisashvili G, Kvesitadze G, Lomidze E, Ugrekhelidze D, Zaalishvili G (2005) Absorption, distribution and transformation of TNT in higher plants. Ecotoxicol Environ Saf (*in Press*)

3. Adler T (1996) Botanical cleanup crews. Sci News 150: 42–43

4. Al-Adil KM, White ER, McChesney MM, Kilgore WW (1974) Translocation of pesticides as affected by plant nutrition. J Agric Food Chem 22: 242–249

5. Allnuff FCT, Ewy R, Renganathan M, Pan RS, Dilley KA (1991) Nigericin and hexylamine effects on localized proton gradients in thylacoids. Biochim Biophys Acta Bioenerg 1059: 28–36

6. Andersen JK (2003) Paraquat and iron exposure as possible synergistic environmental risk factors in Parkinson's disease. Neurotox Res 5: 307–313

7. Anderson TA, Walton BT (1995) Comparative fate of ^{14}C trichloroethylene in the root zone of plants from a former solvent disposal site. Environ Toxicol Chem 14: 2041–2047

8. Anderson TA, Kruger EL, Coats JR (1994) Enhanced degradation of a mixture of three herbicides in the rhizosphere of herbicide-tolerant plants. Chemosphere 28: 1551–1557

9. Andreopoulos-Renaud U, Glas J, Falgoux D, Schiedecker D (1975). Absorption de polyethyleneglycol par de jeunes plantes de haricot et de cotonnier. Dosage par chromatographie en phase gaze use sur exsudat de tige. Cr Acad Sci D 280: 2333–2340

10. Andrews LS, Snyder R (1991) Toxic effects of solvents and vapors. In: Amdur MO, Doull J, Klaassen CD (eds) Cassarett and Doull's toxicology. 4th edn. McGraw Hill, New York, pp 693–694

11. Angermaier L, Simon H (1983) On nitroaryl reductase activities in several *Clostridia*. Hoppe-Seylers Z Physiol Chem 364: 1653–1663

12. Applebury ML, Coleman JE (1969) *Escherichia coli* alkaline phosphatase. J Biol Chem 244: 308–318

13. Aprill W, Sims RC (1990) Evaluation of the use of prairie grasses for stimulating polycyclic aromatic hydrocarbon treatment in soil. Chemosphere 20: 253–265

14. Archakov AI (1983) Oxygenases of biological membranes (in Russian). XXXVII Bakhovskie Chteniya. Nauka, Moscow, pp 1–26
15. Archer TE, Stokes JD, Bringhurst RS (1977) Fate of carbofuran and its metabolites in strawberries in the environment. J Agric Food Chem 25: 536–541
16. Arziani B, Ugrekhelidze D, Mithaishvili T (1983) Detoxification of 2,4-dinitrophenol in plants (in Russian). Fiziol Rast (Moscow) 30: 1040–1042
17. Atkinson MJ (1987) Alkaline phosphatase activity of coral reef benthos. Coral Reefs 6: 59–62
18. ATSDR (1989) Toxicological profile for vinyl chloride. Prepared by Syracuse Research Corporation. Contract No 68-C8-0004. ATSDR/TP-88-25
19. ATSDR (2000) Toxicological profile for arsenic. Final Report. US Department of Health and Human Services, Public Health Service. NTIS Accession No PB2000-108021 Atlanta
20. Audley BG (1979) Structure and properties of 2-chloroethylphosphonic acid (ethephon) metabolite from *Hevea brasiliensis* bark. Phytochemistry 18: 53–60
21. Banerjee HN, Verma M, Hou LH, Ashraf M, Dutta SK (1999) Cytotoxicity of TNT and its metabolites. Yale J Biol Med 72: 1–4
22. Bañuelos GS, Ajwa HA, Mackey B, Wu LL, Cook C, Akohoue S, Zambrzuski S (1997) Evaluation of different plant species used for phytoremediation of high soil selenium. J Environ Qual 26: 639–646
23. Barnes JD, Eamus D, Davison AW, Ro-Poulsen H, Mortensen L (1990) Persistent effects of ozone on needle water loss and wettability in Norway spruce. Environ Pollut 63: 345–363
24. Bataynen N, Kopacz SI, Lee CP (1986) The modes of action of long chain alkali compounds on the respiratory chain-linked energy transducing system in submitochondrial particles. Arch Biochem Biophys 250: 476–487
25. Baur JR, Bovey RW, Riley I (1974) Effect of pH on foliar uptake of 2,4,5-T-1-^{14}C. Weed Sci 22: 481–489
26. Bear J (1979) Hydraulics of groundwater. McGraw-Hill, New York
27. Bellin CA, O'Connor GA (1990) Plant uptake of pentachlorophenol from sludge-amended soils. J Environ Qual 19: 598–602
28. Berenbaum MR, Zangerl AR (1996) Physical-chemical diversity: adaptation or random variation? Rec Adv Phytochem 30: 1–12
29. Berg VS (1989) Leaf cuticles as potential markers of air pollutant exposure in trees. In: Biologic markers of air pollution stress and damage in forests. Natl Acad Press, Washington, pp 333–340
30. Best EPH, Zappi ME, Fredrickson HL, Sprecher SL, Larson SL, Ochman M (1997) Screening of aquatic and wetland plant species for the phytoremediation of explosives-contaminated groundwater from the Iowa Army Ammunition Plant. Ann NY Acad Sci 829: 179–194
31. Best EP, Sprecher SL, Larson SL, Fredrickson HL, Bader DF (1999). Environmental behavior of explosives in groundwater from the Milan Army Ammunition Plant in aquatic and wetland plant treatments. Removal, mass balances and fate in groundwater of TNT and RDX. Chemosphere 38: 3383–3396

32. Best EPH, Sprecher SL, Larson SL, Fredrickson HL, Bader DF (1999) Environmental behavior and fate of explosives from groundwater from the Milan Army Ammunition Plant in aquatic and wetland plant treatments. Uptake and fate of TNT and RDX in plants. Chemosphere 39: 2057–2072

33. Best EPH, Kvesitadze G, Khatisashvili G, Sadunishvili T (2005) Plant processes important for the transformation and degradation of explosives contaminants. Z Naturforsch C Biosci 60c: 340–348

34. Betsiashvili M, Sadunishvili T, Amashukeli N, Tsulukidze N, Shapovalova N, Dzamukashvili N, Nutsubidze N (2004) Effect of aromatic hydrocarbons on main metabolic and energetic enzymes in maize, ryegrass and kidney bean seedlings. Bull Georg Acad Sci 170: 172–174

35. Beyer EMJ, Duffy MJ, Hay JV, Schlueter DD (1988) Sulfuronureas. In: Kearney PC, Kaufman DD (eds) Herbicides: chemistry, degradation and mode of action. Dekker, New York Basel, vol 3, pp 117–189

36. Bhadra R, Wayment DG, Hughes JB, Shanks JV (1999) Confirmation of conjugation processes during TNT metabolism by axenic plant roots. Environ Sci Technol 33: 446–452

37. Bieseler B, Fedtke C, Neuefeind T, Etzel W, Prade L, Reinemer P (1997) Maize selectivity of FOE 5043: Degradation of active ingredient by glutathione-S-transferases. Pflanzenschutz-Nachrichten Bayer 50: 117–140.

38. Boden TA, Kaiser DP, Sepanski RJ, Stoss FW (1994) Trends '93: a compendium of data on global change. Carbon dioxide. Information Analysis Center Communications 20: 810

39. Bokern M, Harms HH (1997) Toxicity and metabolism of 4-n-nonylphenol in cell suspension cultures of different plant species. Environ Sci Technol 31: 1849–1854

40. Bond JA, Bradley BP (1997) Resistance to malathion in heat-shocked *Daphnia magna*. Environ Toxicol Chem 16: 705–712

41. Brazier M, Cole DJ, Edwards R (2002) O-Glucosyltransferase activities toward phenolic natural products and xenobiotics in wheat and herbicide-resistant and herbicide-susceptible black-grass (*Alopecurus myosuroides*). Phytochemistry 59: 149–156

42. Breaux EJ (1987) Initial metabolism of acetochlor in tolerant and susceptible seedlings. Weed Sci 35: 463–468

43. Brewer PE, Wilson RE (1975) Dichloromethane: variability in penetration and resulting effects on seed germination and CO_2 evolution. Bot Gaz (Chikago) 136: 216–218

44. Briggs GG, Bromilow RH, Evans AA (1982) Relationships between lipophilicity and root uptake and translocation of non-ionized chemicals by barley. Pestic Sci 13: 495–504

45. Bromilow RH, Chamberlain K, Evans AA (1990) Physicochemical aspects of phloem translocation of herbicides. Weed Sci 38: 305–314

46. Brown HM, Neighbors SM (1987) Soybean metabolism of chlorimuron ethyl: physiological basis for soybean selectivity. Pestic Biochem Physiol 29: 112–120

47. Brown HM, Wittenbach VA, Forney DR, Strachan SD (1990) Basis for soybean tolerance to thifensulfuron methyl. Pestic Biochem Physiol 37: 303–313
48. Brown SL, Chaney RL, Angle JS, Baker AJM (1994) Phytoremediation potential of *Thlaspi caerulescens* and bladder campion for zinc and cadmium contaminated soil. J Environ Qual 23: 1151–1157
49. Brudenell AJP, Baker DA, Grayson BT (1995) Phloem mobility of xenobiotics, tabular review of physicochemical properties governing output of the Kleier model. J Plant Growth Regul 16: 215–231
50. Brudenell AJP, Griffiths H, Rossiter JT, Baker DA (1999) The phloem mobility of glucosinolates. J Exp Bot 50: 745–756
51. Bryant DW, McCalla DR, Leelsma M, Laneuville P (1981) Type I nitroreductases of *Escherichia coli*. Can J Microbiol 27: 81–86
52. Buadze O, Kvesitadze G (1997) Effect of low-molecular-weight alkanes on the cell photosynthetic apparatus. Ecotoxicol Envoron Saf 38: 36–44
53. Buadze O, Kakhaya M, Zaalishvili G (1979) The influence of the lowest alkanes on chloroplast ultrastructure of some plants (in Russian). In: Works of Session on Defence of Environment. Tbilisi, pp 15–16
54. Buadze O, Durmishidze S, Kakhaya M, Katsitadze K, Apakidze A (1985) The electron microscopic study of some questions of phenoxyacetic acid movement, localization and utilization in some plants (in Russian). Proc Georg Acad Sci Biol Ser 11: 311–318
55. Buadze O, Lomidze E, Kakhaya M, Gagnidze L (1986) The uptake and distribution of radioactive label of 1-12-^{14}C-benzidine in plant cell (in Russian). In: Works of Conf of Uzbekistan Biochemists, Tashkent, pp 181–182
56. Buadze O, Sadunishvili T, Kvesitadze G (1998) The effect of 1,2-benzanthracene and 3,4-benzpyrene on the ultrastructure on maize cells. Int Biodeterior Biodegrad 41: 119–125
57. Bucker J, Guderian R (1994) Accumulation of myoinositol in *Populus* as a possible indication of membrane disintegration due to air-pollution. J Plant Physiol 144: 121–129
58. Bukovac MT, Petracek PD, Fader RG, Morse RD (1990) Sorption of organic compounds by plant cuticles. Weed Sci 38: 289–298
59. Bunge M, Adrian L, Kraus A, Opel M, Lorenz WG, Andreesen JR, Görisch H, Lechner U (2003) Reductive dehalogenation of chlorinated dioxins by an anaerobic bacterium. Nature 421: 357–360
60. Burken JG (2003) Uptake and metabolism of organic compounds: green liver model. In: McCutcheon SC, Schnoor JL (eds) Phytoremediation. Transformation and control of contaminants. Wiley-Interscience, Hoboken, New Jersey, pp 59–84
61. Burken JG, Schnoor JL (1998) Predictive relationships for uptake of organic contaminants by hybrid poplar trees. Environ Sci Technol 32: 3379–3385
62. Burrows WJ, Leworthy DP (1976) Metabolism of N,N-diphenylurea by cytokinin-dependent tobacco callus identification of the glucoside. Biochem Biophys Res Comm 70: 1109–1117

63. Burt ME, Corbin FT (1978) Uptake, translocation and metabolism of pro-pham by wheat *(Triticum aestivum)*, sugarbeet *(Beta vulgaris)*, and alfalfa *(Medicago sativa)*. Weed Sci 26: 296–302

64. Bylund JE, Dyer JK, Feely DE, Martin EL (1990) Alkaline and acid phos-phatases from the extensively halotolerant bacterium *Halomonas elongata*. Curr Microbiol 20: 125–131

65. Cabanne F, Giallardon P, Scalla R (1985) Phytotoxicity and metabolism of chlorotoluron in two wheat varieties. Pestic Biochem Physiol 23: 212–220

66. Carringer RD, Rieck CE, Bush LP (1978) Metabolism of EPTC in corn *(Zea mays)*. Weed Sci 26: 157–163

67. Cash GG (1998) Prediction of chemical toxicity to aquatic microorganism: ECOSAR vs. Microtox assay. Environ Toxicol Water Qual 132: 211–216

68. Cassagne C, Lessire R (1975) Studies on alkane biosynthesis in epidermis of *Allium porrum* L. leaves. 4. Wax movement into and out of the epidermal cells. Plant Sci Lett 5: 261–266

69. Castro S, Davis LC, Erickson LE (2001) Plant-enhanced remediation of gly-col-based aircraft deicing fluids. Pract Period Hazard. Toxic Radioact Waste Manag 5: 141–152

70. Cataldo DA, Harvey S, Fellows RJ, Bean RM, McVeety BD (1989) An evaluation of the environmental fate of munitions material (TNT, RDX) in soil and plant systems: TNT. US DOE Contract 90-012748, Pacific NW Laboratories, Richland

71. Cernicharo J, Heras AM, Tielens AGGM, Pardo JR, Herpin F, Guélin M, Waters LBFM (2001) Infrared space observatory's discovery of C_4H_2, C_6H_2, and benzene in CRL 618. Astrophys 546: L123–L126

72. Chamberlain K, Patel S, Bromilow RH (1998) Uptake by roots and translo-cation to shoots of two morpholine fungicides in barley. Pestic Sci 54: 1–7

73. Chandler JM, Basler E, Santelman PW (1974) Uptake and translocation of alachlor in soybean and wheat. Weed Sci 22: 253–259

74. Chang FY, Vanden Born WH (1971) Dicamba uptake, translocation, me-tabolism and selectivity. Weed Sci 19: 113–122

75. Chapple C (1998) Molecular-genetic analysis of plant cytochrome P450-dependent monooxygenases. Annu Rev Plant Physiol Plant Mol Biol 49: 311–343

76. Chekol T, Vough LR, Chaney RL (2002) Plant-soil-contaminant specificity and phytoremediation of organic contaminants. Int J Phytoremediation 4: 17–26

77. Chen Y, Lucas PW, Wellburn AR (1991) Relationship between foliar injury and changes in antioxidant levels in Red and Norway spruce exposed to acidic mists. Environ Pollut 69: 1–15

78. Cheng TC, Harvey SP, Chen GL (1996) Cloning and expression of a gene encoding a bacterial enzyme for decontamination of organophosphorous nerve agents and nucleotide sequence of the enzyme. Appl Environ Micro-biol 62: 1636–1641

79. Chkanikov DI (1985) Metabolism of 2,4-D in plants (in Russian). Uspekhi Sovremennoi Biologii 99: 212–225

80. Chkanikov DI, Makeev AM, Pavlova NN, Artemenko EN, Dubovoi VP (1976) The role of 2,4-D metabolism in plant resistance to this herbicide (in Russian). Agrokhimiya 2: 127–132

81. Chollet JF, Delétage C, Faucher M, Miginiac L, Bonnemain JL (1997) Synthesis and structure-activity relationships of some pesticides with an alpha-amino acid function. Biochim Biophys Acta 1336: 331–341

82. Chow PNP (1970) Absorption and dessipation of TCA by wheat and oats. Weed Sci 18: 429–438

83. Chrikishvili D, Ugrekhelidze, D, Mithaishvili T (1977) Products of phenol conjugation in maize (in Russian). Bull Georg Acad Sci 88: 173–176

84. Chrikishvili D, Sadunishvili T, Zaalishvili G (2005) Benzoic acid transformation via conjugation with peptides and final fate of conjugates in higher plants. Ecotoxicol Environ Saf (*in Press*)

85. Cohen SM (2001) Lead poisoning: a summary of treatment and prevention. Pediatr Nurs 27: 125–130

86. Cole DJ, Owen WJ (1987) Metabolism of metalaxyl in cell suspension cultures of *Lactuca sativa* L. and *Vitis vinifera* L. Pestic Biochem Physiol 28: 354–361

87. Coleman JOD, Mechteld MA, Kalff B, Davies TGE (1997) Detoxification of xenobiotics in plants: chemical modification and vacuolar compartmentation. Trends Plant Sci 2: 144–151

88. Collins PJ, Dobson ADW (1997) Regulation of laccase gene transcription in *Trametes versicolor*. Appl Environ Microbiol 63: 3444–3450

89. Colombo JC, Cabello MN, Arambarri AM (1996) Biodegradation of aliphatic and aromatic hydrocarbons by natural soil microflora and pure culture of imperfect and ligninolytic fungi. Environ Pollut 94: 355–362

90. Commoner B (1994) The political history of dioxin. Keynote address at the 2nd Citizens Conf on Dioxin. St. Louis, http://www.greens.org/s-r/078/07-03.html

91. Conger RM, Portier R (1997) Phytoremediation experimentation with the herbicide bentazon. Remed Spring 7: 19–37

92. Connelly JA, Johnson MD, Gronwald JW, Wyse DL (1988) Bentazon metabolism in tolerant and susceptible soybean *(Glycine max)* genotypes. Weed Sci 36: 417–423

93. Coupland D, Lutman PJW, Heath C (1990) Uptake, translocation, and metabolism of mecoprop in a sensitive and a resistant biotype of *Stellaria media*. Pestic Biochem Physiol 36: 61–67

94. Csintalan Z, Tuba Z (1992) The effect of pollution on the physiological processes in plants. In: Kovács M (ed) Biological indicators in environmental protection. Ellis Horwood, New York

95. Cummins I, Edwards R (2004) Purification and cloning of an esterase from the weed black-grass (*Alopecurus myosuroides*), which bioactivates aryloxy-phenoxypropionate herbicides. Plant J 39: 894–904

96. Cummins I, Burnet N, Edwards R (2001) Biochemical characterisation of esterases active in hydrolysing xenobiotics in wheat and competing weeds. Physiol Plant 113: 477–485

97. Cunningham SD, Shann JR, Crowley DE, Anderson TA (1997) Phytoreme-diation of contaminated water and soil. In: Kruger EL, Anderson TA, Coats JR (eds) Phytoremediation of soil and water contaminants. ACS Symp Ser No 664. Am Chem Soc, Washington

98. Curfs DM, Beckers L, Godschalk RW, Gijbels MJ, van Schooten FJ (2003) Modulation of plasma lipid levels affects benzo[a]pyrene-induced DNA damage in tissues of two hyperlipidemic mouse models. Environ Mol Mutagen 42: 243–249

99. Davidonis GH, Hamilton RH, Mumma RO (1978) Metabolism of 2,4-dichlorophenoxyacetic acid in soybean root callus and differentiated soybean root cultures as a function of concentration and tissue age. Plant Physiol 62: 80–86

100. Davies HM, Merydlth A, Mende-Mueller L, Aapola A (1990) Metabolic detoxification of phenmedipham in leaf tissue of tolerant and susceptible species. Weed Sci 38: 206–214

101. Davis LC, Muralidharan N, Visser VP, Chaffin C, Fateley WG, Erickson LE, Hammaker RM (1994) Alfalfa plants and associated microorganisms promote biodegradation rather than volatilization of organic substances from groundwater. In: Bioremediation through rhizosphere technology. Am Chem Soc, Washington, pp 112–122

102. Davis WT (2000) Air pollution engineering manual. 2nd edn. Wiley, New York

103. Day BE (1952) The absorption and translocation of 2,4-dichlorophenoxy-acetic acid by bean plants. Plant Physiol 27: 143–153

104. Dec J, Bollag JM (1994) Use of plant material for the decontamination of water polluted with phenols. Biotechnol Bioeng 44: 1132–1139

105. Delétage-Grandon C, Chollet JF, Faucher M, Rocher F, Komor E, Bonne-main JL (2001) Carrier-mediated uptake and phloem system of a 350-Dalton chlorinated xenobiotic with an α-amino acid function. Plant Physiol 125: 1620–1632

106. Demerjian KL, Kerr JA, Calvert JG (1974) The mechanism of photochemical smog formation. In: Pitts JN, Metcalf RL (eds) Advances in environmental science and technology. Wiley, New York, pp 1–262

107. DeRidder BP, Dixon DP, Beussman DJ, Edwards R, Goldsbrough PB (2002) Induction of glutathione S-transferases in Arabidopsis by herbicide safeners. Plant Physiol 130: 1497–1505

108. Devdariani T (1988) Biotransformation of cancerogenic polycyclic aromatic hydrocarbons in plants (in Russian). In: Durmishidze S (ed) Biotransformation of xenobiotics in plants. Metsniereba, Tbilisi, pp 79–162

109. Devdariani T, Durmishidze S (1983) Isolation and identification of the main benzo(a)pyrene oxidation products in plants (in Russian). In: Durmshidze S (ed) Methods of biochemical studies of plans. Metsniereba, Tbilisi, pp 101–124

110. Devdariani T, Kavtaradze L (1979) Study of absorption and transformation of benz[a]anthracene by plant cells in sterile conditions. In: Durmishidze S

(ed) Metabolism of chemical pollutants of biosphere in plants (in Russian). Metsniereba, Tbilisi, pp 92–97

111. Devdariani T, Kavtaradze L, Kvartskhava L (1979) Uptake of benz[a]anthracene-9-^{14}C by roots of annual plants (in Russian). In: Durmishidze S (ed.) Plants and chemical carcinogenics. Metsniereba, Tbilisi, pp 90–95

112. Devine MD, Hall LM (1990) Implication of sucrose transport mechanism for the translocation of herbicides. Weed Sci 38: 299–304

113. Dexter AG, Slife FW, Butler HS (1971) Detoxification of 2,4-D by several plant species. Weed Sci 19: 721–729

114. Di Toro DM, Hellweger FL (1999) Long-range transport and deposition: the role of Henry's law constant. http://www.cefic.be/icca/pops/en/di990527. pdf# search='Henry's%20constant'

115. Didierjean L, Gondet L, Perkins R, Lau S-MC, Schaller H, O'Keefe DP, Werck-Reichhart D (2002) Engineering herbicide metabolism in tobacco and *Arabidopsis* with CYP76B1, a cytochrome P450 enzyme from Jerusalem artichoke. Plant Physiol 130: 179–189

116. Dietz AC, Schnoor JL (2001) Advances in phytoremediation. Environ Health Perspectives 109 Suppl 1: 163–168

117. Dixon DP, Lapthorn A, Edwards R (2002) Plant glutathione transferases. Genome Biol 3: 3004.1–3004.10

118. Dołowy K (2001) Environmental toxins and ion channels. Cell Mol Biol Lett 6: 343–347

119. Domir SC (1978) Translocation and metabolism of injected maleic hydrazide in silver maple and American sycamore seedlings. Physiol Plant 42: 387–396

120. Donnelly PK, Hegde RS, Fletcher JS (1994) Growth of PCB-degrading bacteria on compounds from photosynthetic plants. Chemosphere 28: 981–988

121. Dörr R (1970) Die Aufnahme von 3,4-Benzpyren durch Pflanzenwurzeln. Landwirtsch Forsch 23: 371–376

122. Doty SL, Shang TQ, Wilson AM, Moore AL, Newman LA, Strand SE, Gordon MP (2003) Metabolism of the soil and groundwater contaminants, ethylene dibromide and trichloroethylene, by the tropical leguminous tree, *Leucaena leucocephala*. Water Res 37: 441–449

123. Doty SL, Strand SE (2004) Metabolism of halogenated hydrocarbons by plant cells. In: Phytoremediation: environmental and molecular biological aspects. OECD workshop, Hungary, Abstr, p 67

124. Dresback K, Choshal D, Goyal A (2001) Phycoremediation of trichloroethylene (TCE) Physiol Mol Biol Plants 7: 117–123

125. Droog FJN, Hooykaas PJJ, Libbenga KR, van der Zaal EJ (1993) Proteins encoded by an auxin-regulated gene family of tobacco share limited but significant homology with glutathione S-transferases and one member indeed shows *in vitro* GST activity. Plant Mol Biol 21: 965–972

126. Dueck TA, Wolting HG, Moet DR, Pasman FJM (1987) Growth and reproduction of *Silene cucubalus* Wib. intermittently exposed to low concentrations of air pollutants zinc and copper. New Phytol 105: 633–646

127. Durmishidze S (1988) Plant biochemistry and environmental protection (in Russian). In: Khachidze O (ed) Biotransformation of xenobiotics in plants. Metsniereba, Tbilisi, pp 4–55

128. Durmishidze S, Ugrekhelidze D (1967) Assimilation and translocation of gaseous hydrocarbons by higher plants. In: 7th Int Congr Biochem. Tokyo, Abstr, p J-302

129. Durmishidze S, Ugrekhelidze D (1968) Absorption and conversion of butane by higher plants (in Russian). Dokladi Akademii Nauk SSSR 182: 214–216

130. Durmishidze S, Ugrekhelidze D (1968). Oxidation of ethane, propane and pentane by higher plants (in Russian). Bull Georg Acad Sci 50: 661–666

131. Durmishidze S, Ugrekhelidze D (1975) Absorption and transformation of methane by plants (in Russian). Fiziol Rast (Moscow) 22: 70–73

132. Durmishidze S, Ugrekhelidze D, Djikiya A, Tsevelidze D (1969) The intermediate products of enzymatic oxidation of benzene and phenol (in Russian). Dokladi Akademii Nauk SSSR 184: 466–469

133. Durmishidze S, Ugrekhelidze D, Djikiya A (1974) Absorption and transformation of benzene by higher plants (in Russian). Fiziologiya i Biochimiya Kulturnikh Rastenii 6: 217–221

134. Durmishidze S, Ugrekhelidze D, Djikiya A (1974) Uptake of benzene by fruits from atmosphere (in Russian). Appl Biochem Microbiol 10: 472–476

135. Durmishidze S, Ugrekhelidze D, Djikiya A (1974) Absorption and transformation of toluene by higher plants (in Russian). Appl Biochem Microbiol 10: 673–676

136. Durmishidze S, Djikiya A, Lomidze E (1979) Uptake and transformation of benzidine by plants in sterile conditions (in Russian). Dokladi Akademii Nauk SSSR 247: 244–247

137. Durmishidze S, Ugrekhelidze D, Djikiya A, Lomidze E (1979) Uptake of benzidine by kidney bean seedlings in sterile conditions (in Russian). In: Slepyan EI (ed) Plants and chemical carcinogens. Nauka, Leningrad, pp 93–95

138. Durmishidze S, Ugrekhelidze D, Kakhniashvili C (1982) Metabolism of phenoxyacetic acids in plants: conjugation products of phenoxyacetic and 2,4-dichlorophenoxyacetic acids with peptides. In: 5th Int Congr Pestic Chem (JUPAC). Kyoto, Japan, Abstr, p Va-2

139. Durmishidze S, Devdariani T, Kakhniashvili C, Buadze O (1988) Biotransformation of xenobiotics in plants (in Russian). Metsniereba, Tbilisi

140. Durst F (1991) Biochemistry and physiology of plant cytochrome P-450. In: Ruckpaul K (ed) Frontiers in biotransformation. Academic-Verlag, Berlin, vol 4, pp 191–232

141. Eckardt NA (2001) Move it on out with MATEs. Plant Cell 13: 1477–1480

142. Edwards NT, Edwards GL, Kelly JM, Taylor GE (1992) Three year growth responses of *Pinus taeda* L. to simulated rain chemistry, soil magnesium status and ozone. Water Air Soil Pollut 63: 105–118

143. Edwards R, Owen WJ (1989) The comparative metabolism of the 5-triazine herbicides atrazine and terbutryne in suspension cultures of potato and wheat. Pestic Biochem Physiol 34: 246–254

144. Eglinton G, Hamilton RJ (1963) The distribution of alkanes. In: Swain T (ed) Chemical plant taxonomy. Natl Acad Press, London New York, pp 187–217

145. Eichler W (1982) Gift in unserer Nahrung. Kilda–Verlag, Greven

146. El-Hawari AM, Hodgson JR, Winston JM, Sawyer MD, Hainje M, Lee CC (1981) Species differences in the disposition and metabolism of 2,4,6-trinitrotoluene as a function of route of administration. Final Report by the Midwest Research Institute, Project No. 4274-B, Kansas City, DAMD17-76-C-6066. AD-A114–025

147. Ellis LBM, Hou BK, Kang W, Wackett LP (2003) The University of Minnesota Biocatalysis/Biodegradation Database: post-genomic data mining. Nucleic Acids Res 31: 262–265

148. Ensley HE, Sharma HA, Barber JT, Polito MA (1997) Metabolism of chlorinated phenols by *Lemna gibba,* duckweed. In: Kruger EL, Anderson TA, Coats JR (eds) Phytoremediation of soil and water contaminants. Am Chem Soc, Washington, pp 238–253

149. EPA (1991) Air quality criteria for carbon monoxide. EPA, Office of Research and Development, Washington, EPA-600/B-90/045F

150. EPA (1997) Sulfur dioxide (SO_2). Summary report. http://www.epa.gov/air/airtrends/sulfur.html

151. EPA (2000) Introduction on phytoremediation EPA/600/R-99/107 www.epa.gov/swertio1/download/remed/ introphyto.pdf

152. EPA (2001) Brownfields technology primer: selecting and using phytoremediation for site cleanup. 542-R-01-006, pp 1–24

153. Epstein E, Lavee Sh (1977) Uptake, translocation, and metabolism of IAA in the olive *(Olea europaea):* Uptake and translocation of [1-^{14}C] IAA in detached Manzanilla olive leaves. J Exp Bot 28: 619–625

154. Epuri V, Sorensen DL (1997) Benzo(a)pyrene and hexachlorobiphenyl contaminated soil: phytoremediation potential. In: Kruger EL, Anderson TA, Coats JR (eds) Phytoremediation of soil and water contaminants. Am Chem Soc, Washington, pp 200–222

155. Ernst W (1985) Accumulation in aquatic organisms. In: Sheehan P, Korte F, Klein W, Bourdeau P (eds) Appraisal of tests to predict the environment behaviour of chemicals. Wiley, New York Chichester, pp 285–332

156. Esteve-Núñez A, Caballero A, Ramos JL (2001) Biological degradation of 2,4,6-trinitrotoluene. Microbiol Mol Biol Rev 65: 335–352

157. Eynard I (1974) Determination on foliage of surfactant solution by a radio-isotope technique. Allionia 17: 131–135

158. Ezra G, Stephenson GR (1985) Comparative metabolism of atrazine and EPTC in proso millet (*Panicum miliaceum* L.) and corn. Pestic Biochem Physiol 24: 207–212

159. Fant FA, de Sloovere A, Matthijse K, Marle C, el Fantroussi S, Werstraete W (2001) The use of amino compounds for binding 2,4,6-trinitrotoluene in water. Environ Pollut 111: 503–507

160. Fay M, Donohue JM, De Rosa C (1999) ATSDR evaluation of health effects of chemicals. VI. Di(2-ethylhexyl)phthalate. Toxicol Ind Health 15: 651–746

161. Fedorov LA (1995) A not announced chemical war in Russia. Politics against Ecology (in Russian). Moscow. http://www.seu.ru/cci/lib/books/chemwar/index.htm
162. Fellenberg G (1990) Chemie der Umweltbelastung. Teubner, Stuttgart
163. Fernandez-Bayon JM, Barnes JD, Ollerenshaw JH, Davison AW (1993) Physiological effects of ozone on cultivars of watermelon (*Citrullus lanatus*) and muskmelon (*Cucumis melo*) widely grown in Spain. Environ Pollut 81: 199–206
164. Ferro AM, Sims RC, Bugbee B (1994) Hycrest crested wheatgrass accelerates the degradation of pentachlorophenol in soil. J Environ Qual 23: 272–279
165. Ferro A, Kennedy J, Doucette W, Nelson S, Jauregui G, McFarland B, Bugbee D (1997) Fate of benzene in soils planted with alfalfa: uptake, volatilization, and degradation. In: Kruger EL, Anderson TA, Coats JR (eds) Phytoremediation of soil and water contaminants. Am Chem Soc, Washington, pp 223–237
166. Feung C, Hamilton RH, Mumma RO (1976) Metabolism of 2,4-dichlorophenoxyacetic acid: 10. Identification of metabolites in rice root callus tissue cultures. J Agric Food Chem 24: 1013–1019
167. Fialho RC, Bucker J (1996) Changes in levels of foliar carbohydrates and myoinositol before premature leaf senescence of *Populus nigra* induced by a mixture of O_3 and SO_2. Can J Bot 74: 965–970
168. Fleeker J, Steen R (1971) Hydroxylation of 2,4-D in several weed species. Weed Sci 19: 507–513
169. Fletcher JS, Hegde RS (1995) Release of phenols by perennial plant roots and their potential importance in bioremediation. Chemosphere 31: 3009–3016
170. Fletcher JS, McFarlane JC, Pfleeger T, Wicliff C (1990) Influence of root exposure concentration on the fate of nitrobenzene in soybean. Chemosphere 20: 513–523
171. Fokin AV, Kolomiets AF (1985) Dioxins – scientific or social problem? (in Russian). Nature (Moscow) 3: 3–15
172. Fonné-Pfister R, Kreuz K (1990) Ring-methyl hydroxylation of chlortoluron by an inducible cytochrome P450-dependent enzyme from maize. Phytochemistry 29: 2793–2796
173. Franke W (1975) Stoffaufnahme durch das Blatt unter besonderer Berücksichtigung der Ektodermen. Bodenkultur 26: 331–340
174. Frear DS, Swanson HR (1972) New metabolites of monuron in excised cotton leaves. Phytochemistry 11: 1919–1923
175. Frear DS, Swanson HR, Mansager ER, Wien RG (1978) Chloramben metabolism in plants: isolation and identification of glucose ester. J Agric Food Chem 26: 1347–1354
176. Frear DS, Swanson HR, Mansager ER (1983) Acifluorfen metabolism in soybean: diphenylether bond cleavage and the formation of homoglutathione, cysteine, and glucose conjugates. Pestic Biochem Physiol 20: 299–310

177. Frear DS, Swanson HR, Mansager ER, Tanaka FS (1983) Metribuzin metabolism in tomato: Isolation and identification of N-glucoside conjugates. Pestic Biochem Physiol 19: 270–281

178. Frear DS, Swanson HR, Mansager ER (1985) Alternate pathways of metribuzin metabolism in soybean: Formation of N-glucoside and homoglutathione conjugates. Pestic Biochem Physiol 23: 56–65

179. French CE, Hosser SJ, Davies GJ, Nicklin S, Bruce NC (1999) Biodegradation of explosives by transgenic plants expressing pentaerythritol tetranitrate reductase. Nat Biotechnol 17: 491–494

180. Führ F, Mittelstaedt W (1974) Versuche mit Polyurethan-Hartschaum auf Basis von ^{14}C-markiertem Diphenylmethan-diisocyanat (MDI) zu Fragen der Mineralisierung und Aufnahme durch Planzen. Ber Kernforschungsanlage Jülich 1062: 30–42

181. Führ F, Sauerbeck D (1967) The uptake of colloidal organic substances by plant roots as shown by experiments with ^{14}C-labelled humus compounds. In: Report FAO/IAEA Meeting, Viena. Pergamon Press, Oxford, pp 73–82

182. Furikawa A (1991) Inhibition of photosynthesis of *Populus euramericana* and *Helianthus annuus* by SO_2, and NO_2 and O_3. Ecol Res 6: 79–86

183. Furini A, Fusco N, Dal Corso G, Micheletto L, Borgato L (2004) Transcription profile of genes induced by cadmium in *Brassica juncea*. In: Phytoremediation: environmental and molecular biological aspects. OECD workshop, Hungary, Abstr, p 50

184. Gao JA, Garrison WA, Mazur C, Hoehamer CF, Wolfe NL (2000) Uptake and phytotransformation of organophosphorous pesticides by axenically cultivated aquatic plants. J Agric Food Chem 48: 6121–6127

185. Garner WY, Menzer RE (1986) Metabolism of N-hydroxymethyl dimethionate in bean plants. Pestic Biochem Physiol 25: 218–232

186. Garrison WA, Mzengung VA, Avants JK, Ellington JJ, Jones WJ, Rennels D, Wolfe NL (2000) Phytodegradation of *p,p'*-DDT ant enantiomers of *p,p'*-DDT. Environ Sci Technol 34: 1663–1670

187. Gaskin JL, Fletcher J (1997) The metabolism of exogenously provided atrazine by the ectomycorrhizal fungus *Hebeloma crustuliniforme* and the host plant *Pinus ponderosa*. In: Kruger EL, Anderson TA, Coats JR (eds) Phytoremediation of soil and water contaminants. Am Chem Soc, Washington, pp 152–160

188. Gatliff EG (1994) Vegetative remediation process offers advantages over pump-and-treat. Remediation. Summer/1994: 343–352

189. Gellatly KS, Moorhead GBG, Duff SMG, Lefebvre DD, Plaxton WC (1994) Purification and characterization of a potato tuber acid phosphatase having significant phosphotyrosine phosphatase activity. Plant Physiol 106: 223–232

190. Geyer H, Scheunert I, Korte F (1985) Relationship between the lipid content of fish and their bioconcentration potential of 1,2,4-trichlorobenzene. Chemosphere 14: 545–555

191. Gilbert ES, Crowley DE (1997) Plant compounds that induce polychlori- nated biphenyl biodegradation by *Arthrobacter* sp. Strain B1b. Appl Environ Microbiol 63: 1933–1938
192. Glanze WD (1996) Mosby Medical Encyclopedia. Revised edn. Mosby, St Louis
193. Godbold DL, Feig R, Cremer-Herms A, Huttermann A (1993) Determination of stress bioindicators in three Norway spruce stands in northern Germany. Water Air Soil Pollut 66: 231–243
194. Gordeziani M, Durmishidze S, Khatisashvili G, Adamia G, Lomidze E (1987) The investigation of biosynthetic and detoxification ability of plant cytochrome P-450 (in Russian). Dokladi Akademii Nauk SSSR 295: 1491– 1493
195. Gordeziani M, Khatisashvili G, Kurashvili M (1991) Distribution of NADPH-cytochrome P-450-reductase in plant cell (in Russian). Bull Georg Acad Sci 143: 321–324
196. Gordeziani M, Khatisashvili G, Kvesitadze G (1991) Free and coupled with xenobiotic hydroxylation oxidation of NADPH (in Russian). Dokladi Akademii Nauk SSSR 320: 467–470
197. Gordeziani M, Khatisashvili G, Ananiashvili T, Varazashvili T, Kurashvili M, Kvesitadze G, Tkhelidze P (1999) Energetic significance of plant monooxygenase individual components participating in xenobiotic degrada- tion. Int Biodeterior Biodegrad 44: 49–54
198. Goyer RA (1996) Toxic effects of metals: mercury. Casarett and Doull's Toxicology: the basic science of poisons, 5th edn. McGraw-Hill, New York
199. Grayson BT, Kleier DA (1990) Phloem mobility of xenobiotics. IV. Model- ing of pesticide movement in plants. Pestic Sci 30: 67–79
200. Greene DW, Bukovac MJ (1977) Foliar penetration of naphthalene-acetic acid enhancement by light and role of stomata. Am J Bot 64: 96–104
201. Groom CA, Halasz A, Paquet L, Morris N, Olivier L, Dubois C, Hawari J. (2002) Accumulation of HMX (octahydro-1,3,5,7-tetranitro-1,3,5,7- tetrazocine) in indigenous and agricultural plants grown in HMX- contaminated anti-tank firing-range soil. Environ Sci Technol 36: 112–118
202. Gross D, Laanio T, Dupius T, Esser HO (1979) The metabolic behaviour of chlorotoluron in wheat and soil. Pestic Biochem Physiol 10: 49–53
203. Guillén F, Martýnez MJ, Muñoz C, Martýnez AT (1997) Quinone redox cy- cling in the ligninolytic fungus *Pleurotus eryngii* leading to extracellular production of superoxide anion radical. Arch Biochem Biophys 339: 190– 199
204. Guillén F, Gómez-Toribio V, Martýnez MJ, Martýnez AT (2000) Production of hydroxyl radical by the synergistic action of fungal laccase and aryl alco- hol oxidase. Arch Biochem Biophys 382: 142–147
205. Gullner G, Komivec N, Rennenberg H (2004) Detoxification of chloro- acetinilide by transgenic poplars. In: Phytoremediation: environmental and molecular biological aspects. OECD workshop, Hungary, Abstr, p 24

206. Guo YL, Lambert GH, Hsu CC, Hsu MM (2004) Yucheng: health effects of prenatal exposure to polychlorinated biphenyls and dibenzofurans. Int Arch Occup Environ Health 77: 153–158

207. Guttes S, Failing K, Neumann K, Kleinstein J, Georgii S, Brunn H (1998) Chlororganic pesticides and polychlorinated biphenyls in breast tissue of women with benign and malignant breast disease. Arch Environ Contam Toxicol 35:140–147

208. Gyulai G, Humpreys M, Bittsánszky A, Skøt K, SzabóZ, Heywood S, Kiss J, Cullner G, Skøt L, Radimszky L, Lovatt A, Lagler L, Abberton M, Roderick H, Rennenberg H, Kömíves T, Heszky L (2004) Cut clone stability and improved phytoextraction of transgenic ^{gshl} poplar clones (*Populus canescens) in vitro*. In: Phytoremediation: environmental and molecular biological aspects. OECD workshop, Hungary, Abstr, p 53

209. Haderlie RC (1980) Absorption and translocation of buthidazole. Weed Sci 28: 352–360

210. Hallahan DL, Lau SM, Harder PA, Smiley DW, Dawson GW, Pickett JA, Christoffersen RF, O'Keefe DP (1994) Cytochrome P-450-catalysed monoterpenoid oxidation in catmint (*Nepeta racemosa*) and avocado (*Persea americana*) evidence for related enzymes with different activities. Biochim Biophys Acta 1201: 94–100

211. Hallmen U (1974) Translocation and complex formation of picloram and 2,4-D in rape and sunflower. Physiol Plant 32: 78–83

212. Han S, Hatzios KK (1991) Uptake, translocation, and metabolism of [^{14}C] pretilachlor in fenclorim-safened and unsafened rice seedlings. Pestic Biochem Physiol 39: 281–290

213. Hannink N, Rosser SJ, French CE, Basran A, Murray JA, Nicklin S, Bruce NC (2001) Phytodetoxification of TNT by transgenic plants expressing a bacterial nitroreductase. Nat Biotechnol 19: 1168–1172

214. Hannink N, Rosser SJ, Bruce NC (2002) Phytoremediaition of explosives. Crit Rev Plant Sci 21: 511–538

215. Hansikova H, Frei E, Anzenbacher P, Stiborova M (1994) Isolation of plant cytochrome P450 and NADPH:cytochrome P450-reductase from tulip bulbs (*Tulipa fosteriana)*. Gen Physiol Biophys 13: 149–169

216. Haque A, Weisgerber, J, Klein W (1977) Absorption, efflux and metabolism of the herbicide (^{14}C) buturon as affected by plant nutrition. J Exp Bot 28: 468–472

217. Harborne JB (1977) Introduction to ecological biochemistry. Natl Acad Press, London New York San Francisco

218. Harms H (1975) Metabolisierung von Benzo(a)pyren in pflanzlichen Zellsuspensionkulturen and Weizenkeimpflanzen. Landbauforsch Völkenrode 25: 83–90

219. Harms H, Dehnen W, Monch W (1977) Benzo[a]pyrene metabolites formed by plant cells. Z Naturforsch 320: 321–326

220. Harms H, Bokern M, Kolb M, Bock C (2003) Transformation of organic contaminants by different plant systems. In: McCutcheon SC, Schnoor JL

(eds) Phytoremediation. Transformation and control of contaminants. Wiley-Interscience, Hoboken, New Jersey, pp 285–316

221. Harner T, Bidleman TF, Jantunen LMM, Mackay D (2001) Soil-air exchange model of persistent pesticides in the US Cotton Belt. Environ Toxicol Chem 20: 1612–1621

222. Harvey SD, Fellows RJ, Cataldo DA, Bean RM (1990) Analysis of 2,4,6-trinitrotoluene and its transformation products in soils and plant tissues by high-performance liquid chromatography. J Chromatogr 518: 361–374

223. Hatzios KK, Penner D (1980) Site of uptake and translocation of ^{14}C-buthidazole in corn (*Zea mays*) and redroot pigweed (*Amaranthus retroflexus*). Weed Sci 28: 285–291

224. Hawf LR, Behrens R (1974) Selectivity factors in the response of plants to 2,4-D. Weed Sci 22: 245–249

225. Haynes CA, Koder RL, Miller A-F, Rodgers DW (2002) Structures of nitroreductase in three states. J Biol Chem 277: 11513–11520

226. Heggestad HE (1991) Origin of Bel-W3, Bel-C and Bel-B tobacco varieties and their use as indicators of ozone. Environ Pollut 74: 264–291

227. Heinonsalo J, Jorgensen KS, Haahtela K, Sen R (2000) Effects of *Pinus sylvestris* root growth and mycorrhizosphere development on bacterial carbon source utilization and hydrocarbon oxidation in forest and petroleum-contaminated soils. Can J Microbiol 46: 451–464

228. Hendrick LW, Meggitt WF, Penner D (1974) Basis for selectivity of phenmedipham and desmodipham on wild mustard, redroot pigweed, and sugar beet. Weed Sci 22: 179–186

229. Henry JR (2000) An overview of the phytoremediation of lead and mercury. A report prepared for the U.S. Environmental Protection Agency Office of Solid Waste and Emergency Response Technology Innovation Office. http://www.cluin.org/download/remed/henry.pdf

230. Hippeli S, Elstner EF (1996) Mechanisms of oxygen activation during plant stress – biochemical effects of air-pollutants. J Plant Physiol 148: 249–257

231. Hoagland RE, Zablotowicz RM, Locke MA (1997) An integrated phytoremediation strategy for chloroacetamide herbicides in soil. In: Kruger EL, Anderson TA, Coats JR (eds) Phytoremediation of soil and water contaminants. Am Chem Soc, Washington, pp 92–105

232. Hodgson E, Goldstein JA (2001) Metabolism of toxicants: phase I reactions and pharmacogenetics. In: Hodgson E, Smart RC (eds) Introduction to biochemical toxicology. 3rd edn, Wiley, USA, pp 67–114

233. Hodgson RH, Hoffer BL (1977) Diphenamid metabolism in pepper and an ozone effect. 1. Absorption, translocation, and the extent of metabolism. Weed Sci 25: 324–330

234. Holoubek I, Korinek P, Seda Z, Schneiderova E, Holoubkova I, Pacl A, Triska J, Cudlin P (2000) The use of mosses and pine needles to detect atmospheric persistent organic pollutants at the local and regional scale. Environ Pollut 109: 283–292

235. Holton TA, Cornish EC (1995) Genetics and biochemistry of anthocyanin biosynthesis. Plant Cell 7: 1071–1083

236. Hoskin FCG, Walker JE, Mello CM (1999) Organophosphorous acid anhydrolase in slime mold, duckweed and mung bean: a continuing search for a physiological role and a natural substrate. Chem Biol Interact 119/120: 399–404

237. Hou FS, Milke MW, Leung DW, MacPherson DJ (2001) Variations in phytoremediation performance with diesel-contaminated soil. Environ Technol 22: 215–222

238. Hsu FC, Kleier DA (1990) Phloem mobility of xenobiotics. III. Sensitivity of unified model to plant parameters and application to patented chemical hybridizing agents. Weed Sci 38: 315–323

239. Hsu FC, Kleier DA, Melander WH (1988) Phloem mobility of xenobiotics. II. Bioassay testing of the unified model. Plant Physiol 86: 811–816

240. Hsu TS, Bartha R (1976) Hydrolyzable and non-hydrolyzable 3,4-dichloroaniline–humus complexes and their respective rates of biodegradation. J Agric Food Chem 24: 118–122

241. Hsu TS, Bartha R (1979) Accelerated mineralization of two organophosphate insecticides in the rhizosphere. Appl Environ Microbiol 37: 36–41

242. Hu C, Huystee RB (1989) Role of carbohydrate moieties in peanut peroxidases. Biochem J 263: 129–135

243. Hughes JB, Shanks JV, Vanderford M, Lauritzen J, Bhadra R (1997) Transformation of TNT by aquatic plants and plant tissue cultures. Environ Sci Technol 31: 266–271

244. Hutber GN, Lord EI, Loughman BC (1978) The metabolic fate of phenoxy acetic acids in higher plants. J Exp Bot 29: 619–624

245. Hutchinson JM, Shapiro R, Sweetser PB (1984) Metabolism of chlorimuron by tolerant broadleaves. Pestic Biochem Physiol 22: 243–247

246. IARC (1987) IARC monographs on the evaluation of carcinogenic risks to humans. Overall Evaluations of Carcinogenicity. IARC, Lyon, France

247. Inoue J, Chamberlain K, Bromilow RH (1998) Physicochemical factors affecting the uptake by roots and translocation to shoots of amine bases in barley. Pestic Sci 54: 8–21

248. IPSC (1997) Nitrogen oxides. Environmental Health Criteria 188. 2nd edn. http://www.inchem.org/documents/ehc/ehc/ehc188.htm#SubSectionNumber: 2.3.4

249. Isaacson P (1986) Uptake and transpiration of 1,2-dibromoethane by leaves. Plant Soil 95: 431–434

250. Isensee AR, Jones GE, Turner BC (1971) Root absorption and translocation of picloram by oats and soybeans. Weed Sci 19: 727–736

251. Jacobson A, Shimabukuro RH (1984) Metabolism of diclofop-methyl in root-treated wheat and oat seedlings. J Agric Food Chem 32: 742–748

252. Janes BE (1974) The effect of molecular size, concentration in nutrient solution, and exposure time on the amount and distribution of polyethylene glycol in pepper plants. Plant Physiol 54: 226–236

253. Jansen EF, Olson AC (1969) Metabolism of carbon-14-labelled benzene and toluene in avocado fruit. Plant Physiol 44: 786–791

254. Jassen DB, Withold B (1992) Aerobic and anaerobic degradation of halogenated aliphatics. In: Sigel H (ed) Metal ions in biological systems. Decker, New York, pp 229–327

255. Jenings JC, Coolbaugh RC, Nakata DA, West CA (1993) Characterization and solubilization of kaurenoic acid hydroxylase from *Gibberella fujikuroi*. Plant Physiol 101: 925–930

256. Jensen KIN, Stephenson GR, Hunt LA (1977) Detoxification of atrazine in three *Gramineae* subfamilies. Weed Sci 25: 212–218

257. Johannes C, Majcherczyk A (2000) Natural mediators in the oxidation of polycyclic aromatic hydrocarbons by laccase mediator systems. Appl Environ Microbiol 66: 524–528

258. Jones DW, Foy CL (1972) Absorption and translocation of bioxone in cotton. Weed Sci 20: 116–124

259. Jonsson S, Baun A (2003) Toxicity of mono- and diesters of *o*-phthalic esters to a crustacean, a green alga, and a bacterium. Environ Toxicol Chem 22: 3037–3043

260. Julkunen-Titto R, Lavola A, Kainulainen P (1995) Does SO_2 fumigation change the chemical defense of woody plants: The effect of short-term SO_2 fumigation on the metabolism of deciduous *Salix myrsinifolia* plants. Water Air Soil Pollut 83: 195–203

261. Kakhniashvili C (1988). Biotransformation of some pesticides in plants (in Russian). In: Durmishidze S (ed) Biotransformation of xenobiotics in plants. Metsniereba, Tbilisi, pp 147–163

262. Kakhniashvili C, Mithaishvili T, Ugrekhelidze D (1979) Degradation of aromatic ring of phenoxyacetic acids in plants (in Russian). In: Durmishidze S (ed) Metabolism of chemical pollutants of biosphere in plants. Metsniereba, Tbilisi, pp 82–91

263. Kasana MS, Lea PJ (1994) Growth-responses of mutants of spring barley to fumigation with SO_2 and NO_2 in combination. New Phytol 126: 629–636

264. Kassel AG, Ghoshal D, Goyal A (2002) Phytoremediation of trichloroethylene using hybrid poplar. Physiol Mol Biol Plants 8: 1–8

265. Keller T, Matyssek R (1990) Limited compensation of ozone stress by potassium in Norway spruce. Environ Pollut 67: 1–23

266. Khatisashvili G, Gordeziani M, Kvesitadze G, Korte F (1997) Plant monooxygenases: Participation in xenobiotic oxidation. Ecotoxicol Environ Saf 36: 118–122

267. Khatisashvili G, Adamia G, Pruidze M, Kurashvili M, Varazashvili T, Ananiashvili T (2003) TNT uptake and nitroreductase activity of plants. In: 8th Int FZK/TNO Conf ConSoil 2003, Gent, Belgium, pp 2482–2485

268. Khatisashvili G, Kvesitadze G, Adamia G, Gagelidze N, Sulamanidze L, Ugrekhelidze D, Zaalishvili G, Ghoghoberidze M, Ramishvili M (2004) Bioremediation of contaminated soils on the former military locations and proving grounds in Georgia. J Biol Phys Chem 4: 162–168

269. Kimbara K, Hashimoto T, Fukuda M, Koana T, Takagi M, Oishi M, Yano K (1989) Cloning and sequencing of two tandem genes involved in the degradation of 2,3-dihydroxybiphenyl to benzoic acid in the polychlorinated bi-

phenyl-degrading soil bacterium *Pseudomonas* sp. strain KKS102. J Bacteriol 171: 2740–2747

270. King MG, Radosevich R (1979) Tanoak (*Lithocarpus densiflorus*) leaf surface characteristics and absorption of triclopyr. Weed Sci 27: 599–605

271. Kleier DA (1994) Phloem mobility of xenobiotics. V. Structural requirements for phloem-systemic pesticides. Pestic Sci 42: 1–11

272. Knuteson SL, Whitwell T, Klaine SJ (2002) Influence of plant age and size on simazine toxicity and uptake. J Environ Qual 31: 2096–2103

273. Kochs G, Grisebach H (1986) Enzymatic synthesis of isoflavonoids. Eur J Biochem 155: 311–318

274. Kolattukudy PE (1980) Cutin suberin and waves. In: Stumpf PK (Ed) The biochemistry of plants: a comprehensive treatise. vol 4. Lipids: Structure and Function. Natl Acad Press, New York, pp 571–645

275. Kőmives T, Gullner G (2004) Phase I xenobiotic metabolic systems in plants. In: Phytoremediation: environmental and molecular biological aspects. OECD workshop, Hungary, Abstr, p 25

276. Konradsen F, Van der Hoek W, Amerasinghe FP, Mutero C, Boelee E (2004) Engineering and malaria control: learning from the past 100 years. Acta Trop 89: 99–108

277. Korte F, Behadir M, Klein W, Lay JP, Parlar H, Sceunert I (1992) Lehrbuch der ökologischen Chemie. Grundlagen und Konzepte fur die ökologische Beurteilung von Chemikalien. Georg Thieme Verlag, Stuttgart New York

278. Korte F, Kvesitadze G, Ugrekhelidze D, Gordeziani M, Khatisashvili G, Buadze O, Zaalishvili G, Coulston F (2000) Review: Organic toxicants and plants. Ecotoxicol Environ Saf 47: 1–26

279. Kouji H, Masuda T, Matsunaka S (1990) Mechanism of herbicidal action and soybean selectivity of AKH-7088, a novel diphenyl ether herbicide. Pestic Biochem Physiol 37: 219–226

280. Kovács M (1992) Herbaceous (flowering) plants. In: Kovács M (ed) Biological indicators in environmental protection. Ellis Horwood, New York

281. Krell HW, Sandermann H (1985) Plant biochemistry of xenobiotics. Purification and properties of a wheat esterase hydrolyzing the plasticizers chemical, bis(2-ethylhexyl)phthalate. Eur J Biochem 143: 57–62

282. Kristich MA, Schwarz OJ (1989) Characterization of ^{14}C-naphthol uptake in excised root segments of clover (*Trifolium pratense* L.) and fescue (*Festuca arundinacea* Screb.). Environ Monit Assess 13: 35–44

283. Kumar P, Moran D (2002) Photochemical smog: mechanism, ill-effects, and control. J TERI Inform Digest Energy Environ 1: 445–456

284. Kumar VN, Subrash C (1990) Influences of SO_2 on seed germination and seedling growth of *Eulalipsis binate* (Rets) Hubbord. Geobios (India) 17: 190–191

285. Kurumata M, Takahashi M, Sakamoto A, Ramos JL, Nepovim A, Vanek T, Hirata T, Morikawa H (2004) Degradation of nitrocompounds by transgenic plants expressing a bacterial nitroreductase gene. In: Phytoremediation: environmental and molecular biological aspects. OECD workshop, Hungary, Abstr, p 54

286. Kutchan TM (1995) Alkaloid biosynthesis – the basis for metabolic engineering of medicinal plants. Plant Cell 7: 1059–1061
287. Kvesitadze GI, Bezborodov AM (2002) Introduction to biotechnology (in Russian). Nauka, Moscow
288. Kvesitadze G, Gordeziani M, Khatisashvili G, Sadunishvili T, Ramsden JJ (2001) Some aspects of the enzymatic basis of phytoremediation. J Biol Phys Chem 1: 49–57
289. Lamoureux GL, Rusnes DG (1989) Propachlor metabolism in soybean plants, excised soybean tissues, and soil. Pestic Biochem Physiol 34: 187–204
290. Lan PT, Main R, Motoyama N, Dauterman WC (1983) Hydrolysis of malathion by rabbit liver oligomeric and monomeric carboxylesterases. Pestic Biochem Physiol 20: 232–237
291. Lao S-H, Loutre C, Brazier M, Coleman JOD, Cole DJ, Edwards R, Theodoulou FL (2003) 3,4-Dichloroaniline is detoxified and exported via different pathways in *Arabidopsis* and soybean. Phytochemistry 63: 653–661
292. Larcher W (1995) Physiological plant ecology. Ecophysiology and stress physiology of functional groups. 3rd edn. Springer, Berlin Heidelberg New York
293. Larson SL, Weiss CA, Escalon BL, Parker D (1999) Classification of explosives transformation products in plant tissue. Environ Toxicol Chem 18: 1270–1276
294. Laurent FMG (1994) Chloroaniline peroxidation by soybean peroxidases. Pestic Sci 40: 25–30
295. Lavy TL (1975) Effects of soil pH and moisture on the direct radioassay of herbicides in soil. Weed Sci 23: 49–58
296. Lawlor DW (1970) Absorption of polyethylene glycols by plants and their effects on plant growth. New Phytol 69: 501–506
297. Lay MM, Casida JE (1976) Dichloroacetamide antidotes enhance thiocarbamate sulfoxide detoxification by elevating corn root glutathione and glutathione S-transferase activity. Pestic Biochem Physiol 6: 442–446
298. Le Baron HM, McFariand JE, Simoneaux BJ, Ebert E (1988) Metolachlor. In: Kearney PC, Kaufman DD (eds) Herbicides: chemistry, degradation, and mode of action. vol 3. Dekker, New York, pp 335–381
299. Leah JM, Worrall TL, Cobb AH (1989) Metabolism of bentazon in soybean and the influence of tetcyclasis, BAS 110 and BAS 111. In: Crop Prot Conf – Weeds. Proc Int Conf. Brighton, vol 2, pp 433–440
300. Leah JM, Worrall TL, Cobb AH (1991) A study of bentazon uptake and metabolism in the presence and the absence of cytochrome P-450 and acetyl-coenzyme A carboxylase inhibitors. Pestic Biochem Physiol 39: 232–239
301. Leavitt JR, Penner D (1979) *In vitro* conjugation of glutathione and other thiols with acetanilide herbicides and EPTC sulfoxide and the action of the herbicide antidote R-25788. J Agric Food Chem 27: 533–536
302. Lederer B, Boger P (2004) A ligand function of glutation S-transfrase. In: Phytoremediation: environmental and molecular biological aspects. OECD workshop, Hungary, Abstr, p 15

303. Leece DR (1978) Foliar absorption in *Prunus domestica* L. I. Nature and development of the surface wax barrier. Aust J Plant Physiol 5: 749–752
304. Leienbach KW, Heeger V, Neuhann H, Barz W (1975) Stoffwechsel und Abbau von Nicotinsäure und ihren Derivaten in pflanzlichen Zellsuspensionkulturen. Planta Med Suppl: 148–153
305. Leroux P, Gredt M (1975) Absorption of methylbenzimidazol-2-yl carbamate (carbendazim) by corn roots. Pestic Biochem Physiol 5: 507–512
306. Letham DS, Summons RE, Entsch B, Gollnow BI, Parker CW, McLeod JK (1978) Glucosylation of cytokinin analogues. Phytochemistry 17: 2053–2059
307. Levi PE, Hodgson E (1992) Metabolism of organophosphorous compounds by the flavin-containing monooxygenase. In: Chambers JE, Levi PE (eds) Organophosphorus: chemistry, fate, and effects. Natl Acad Press, San Diego, pp 141–145
308. Lewer P, Owen WJ (1990) Selective action of the herbicide triclopyr. Pestic Biochem Physiol 36: 187–200
309. Li ZS, Szczypka M, Lu YP, Thiele DJ, Rea PA (1996) The yeast cadmium factor protein (YCF1) is a vacuolar glutathione S-conjugate pump. J Biol Chem 271: 6509–6517
310. Libbert E (1974) Lehrbuch der Planzenphysiologie. VEB Gustav Fischer Verlag, Jena
311. Lin Q, Mendelssohn IA (1997) Phytoremediation for oil spill cleanup and habitat restoration in Louisiana coastal marches: effects of march plant species and fertilizer. Technical Report Ser 97-006. Wetland Biogeochemistry Institute, Louisiana State University
312. Lingle SE, Suttle JC (1985) A model system for the study of 2,4-D translocation in leaf of spurge. Can J Plant Sci 65: 369–377
313. Liss PS, Slater PG (1974) Flux of gases across the air-sea interface. Nature 247: 181–184
314. Liu LC, Shimabukuro RH, Nalevaja JD (1978) Diuron metabolism in two sugarcane *(Saccharum officinarum)* cultivars. Weed Sci 26: 642–648
315. Löffler D, Kruse H (1985) Zur Toxikologie des Cadmiums. Schriftenreihe der Untersuchungsstelle für Umwelttoxikologie des Landes Schleswig-Holstein. Kiel, Heft 12
316. Loh A, West SD, Macy TD (1978) Gas chromatographic analysis of tebuthiuron and its grass, sugarcane and sugarcane by-products. J Agric Food Chem 26: 410–415
317. Long JW, Basler E (1974) Patterns of phenoxy herbicide translocation in bean seedlings. Weed Sci 22: 18–24
318. Loutre C, Dixon DP, Brazier M, Slater M, Cole DJ, Edwards R (2003) Isolation of a glucosyltransferase from *Arabidopsis thaliana* active in the metabolism of the persistent pollutant 3,4-dichloroaniline. Plant J 34: 485–493
319. Lu YP, Li ZS, Rea PA (1997) AtMPR1 gene of *Arabidopsis* encodes a glutathione S-conjugate pump: isolation and functional definition of a plant ATP-binding cassette transporter gene. Proc Natl Acad Sci USA 94: 8243–8248

320. Macek T, Macková M, Pavlíková D, Száková J, Truska M, Singh-Cundy A, Kotraba P, Yancey N, Scouten WH (2002) Accumulation of cadmium by transgenic tobacco. Acta Biotechnologica 22: 101–106

321. Macek T, Sura M, Francova K, Chrastilova Z, Pavlíková D, Sylvestre M, Szekeres M, Scouten WH, Kotraba P, Macková M (2004) Approaches using GM plants for the removal of xenobitics (Cd, Ni, PCB) including experiments in real contaminated soils. In: Phytoremediation: environmental and molecular biological aspects. OECD workshop, Hungary, Abstr, p 27

322. Machler F, Wasescha MR, Krieg F, Oertli JJ (1995) Damage by ozone and protection by ascorbic acid in barley leaves. J Plant Physiol 147: 469–473

323. Mackay D, Paterson S (1981) Calculating fugacity. Environ Sci Technol 15: 1006–1014

324. Mackova M, Macek T, Ocenaskove J, Burkhard J, Pazlarova DK (1997) Biodegradation of PCBs by plant cells. Int Biodeterior Biodegrad 39: 317–325

325. Manninen S, Huttunen S (1995) Scots pine needles as bioindicators of sulphur deposition. Can J For Res 25: 1559–1569

326. Marrs KA (1996) The function and regulation of glutathione S-transferases in plants. In: Annu Rev Plant Physiol Plant Mol Biol 47: 127–158

327. Martinova E (1993) An ATP-dependent glutathione-S-conjugate "export" pump in the vacuolar membrane of plants. Nature 364: 247–249

328. Mathur AK, Jhamaria SL (1975) Translocation studies of two systemic fungicides. Pesticides 9: 40–46

329. Mayer A (1987) Polyphenoloxidases in plants-recent progress. Phytochemistry 26: 11–20

330. McComb AJ, McComb JA (1978) Differences between plant species in their ability to utilize substituted phenoxybutyric acids as a source of auxin for tissue culture growth. Plant Sci Lett 11: 227

331. McFadden JJ, Frear DS, Mansager ER (1989) Aryl hydroxylation of diclofop by a cytochrome P-450 dependent monooxygenase from wheat. Pestic Biochem Physiol 34: 92–100

332. McFadden JJ, Gronwald JW, Eberlein CV (1990) In vitro hydroxylation of benzaton by microsomes from naphthalic anhydride-treated corn shoots. Biochem Biophys Res Comm 168: 206–211

333. McFarlane JC, Nolt C, Wickliff C, Pfleeger T, Shimabuku R, McDowell M (1987) The uptake, distribution and metabolism of four organic chemicals by soybean plants and barley roots. Environ Toxicol Chem 6: 847–856

334. Mcleod AR (1995) An open-air system for exposure of young forest trees to sulfur-dioxide and ozone. Plant Cell Environ 18: 215–225

335. Medina VF, Larson SL, Agwaramgbo L, Perez W. (2002) Treatment of munitions in soils using phytoslurries. Int J Phytoremediation 4: 143–156

336. Merbach W, Schilling G (1977) Ursachen der Unempfindlichkeit von Beta vulgaris L. gegenüber Pyrazon, Phenmedipham und Benzthiazuron. Biochem Physiol Pflanz 171: 187–190

337. Meskhi A (1973) Some peculiarities of correlation of structure and biologically active phenolic compounds in the culture of plant tissue sterile seed-

lings (in Russian). In: Durmishidze (ed) Plant biochemistry. Metsniereba, Tbilisi, pp 255–264

338. Middleton W, Jarvis BC, Booth A (1978) The effect of ethanol on rooting and carbohydrate metabolism in stem cuttings of *Phaseolus aureus* Roxb. New Phytol 81: 2790–2797

339. Miller D (1978) Models for total transport. In: Butler GC (ed) Principles of ecotoxicology. Wiley, New York Chichester, pp 71–90

340. Minshall WH, Sample K, Robinson JR (1977) The effect of urea on atrazine uptake from soil. Weed Sci 25: 460–469

341. Mithaishvili T, Kakhniashvili C, Ugrekhelidze D (1979) Products of phenoxyacetic acid conjugation in annual plants (in Russian). In: Durmishidze S (ed) Metabolism of chemical pollutants of biosphere in plants. Metsniereba, Tbilisi, pp 73–81

342. Mithaishvili T, Scalla R, Ugrekhelidze D, Tsereteli B, Sadunishvili T, Kvesitadze G (2005) Transformation of aromatic compounds in plants grown in aseptic conditions. Z Naturforsch C Biosci 60c: 97–102

343. Mitsevich EV, Mitsevich IP, Pereligin VV, Do Ngok Lan, Nguen Tkhau Hoiya (2000) Microorganisms as potential indicators of integral dioxin defoliant pollutions of soils. Appl Biochem Microbiol 36: 582–588

344. Mohn W, Tiedje JM (1992) Microbial reductive dehalogenation. Microbiol Rev 56: 482–507

345. Morant M, Bak S, Moller BL, Werck-Reichhart D (2003) Plant cytochromes P450: tools for pharmacology, plant protection and phytoremediation. Curr Opin Biotechnol 2: 151–162

346. Moriarty F (1985) Bioaccumulation in terrestrial food chains. In: Sheehan P, Korte F, Klein W, Bourdeau P (eds) Appraisal of tests to predict the environment behaviour of chemicals. Wiley, New York Chichester, pp 257–284

347. Morita N, Nakazato H, Okuyama H, Kim Y, Thompson GA (1996) Evidence for a glycosylinositolphospholipid-anchored alkaline phosphatase in the aquatic plant *Spirodela oligorrhiza*. Biochim Biophys Acta 1290: 52–53

348. Mougin C, Cabanne F, Canivenc M-C, Scalla R (1990) Hydroxylation and N-demethylation of chlortoluron by wheat microsomal enzymes. Plant Sci 66: 195–203

349. Müller H (1976) Aufnahme von 3,4-Benzpyren durch Nahrungspflanzen aus künstlich angereicherten Substraten. Z Pflanzenernähr Bodenkd 6: 685–690

350. Nanda Kumar PBA, Dushenkov V, Motto H, Raskin I (1995) Phytoextraction: the use of plants to remove heavy metals from soils. Environ Sci Technol 29: 1232–1238

351. Natural Resource Council Committee on Oil in the Sea (2003) Global marine oil pollution information gateway. http://oils.gpa.unep.org/facts/ sources.htm

352. Nayaranan M, Erickson LE, Davis LC (1999) Simple plant-based design strategy for volatile organic pollutants. Environ Prog 18: 231–242

353. Neighbors SM, Privalle LS (1990) Metabolism of primisulfuron by barnyard grass. Pestic Biochem Physiol 37: 145–153

354. Newman LA, Strand SE, Choe N, Duffy J, Ekuan G. Ruszaj M, Shurtleff BB, Wilmoth J, Heilman P, Gordon MP (1997) Uptake and biotransforma-

tion of trichlorethylene by hybrid poplars. Environ Sci Technol 31: 1062–1067

355. Newman LA, Wang X, Miuznieks IA, Eukan G, Ruszaj M, Cortelucci R, Dormoes, D Karsig G, Newman T, Crampton RS, Hashmonay RA, Yost MG, Heilman PE, Duffy J, Gordon MP, Strand SE (1999) Remediation of trichloroethylene in an artificial aquifer with trees: a controlled field study. Environ Sci Techn 33: 2257–2265

356. Nicolas J, Cheynier V, Fleuriet A, Rouet-Mayer MA (1993) Polyphenols and enzymatic browning. In: Scalbert A (ed) Polyphenolic phenomena. INRA edn. Paris, pp 99–107

357. Niku-Paavola ML, Viikari L (2000) Enzymatic oxidation of alkenes. J Mol Cat 10: 435–444

358. Nordberg G, Jin T, Leffler P, Svensson M, Zhou1 T, Nordberg M (2000) Metallothioneins and diseases with special reference to cadmium poisoning. Analusis 28: 396

359. Novojhilov KV (1977) Problems of dynamics and metabolism of insecticides in plants connected with their rational application (in Russian). Trudy Vsesoyuznogo Instituta Zernovykh Rastenii, Leningrad, pp 5–16

360. O'Brien C (2002) Success for carbon dioxide burial. New Scientist.com news service 10 September. http://www.newscientist.com/news/news.jsp?id=ns 99992779

361. O'Connel KM, Breaux EJ, Fraley RT (1988) Different rates of metabolism of two chloroacetanilide herbicides in pioneer 3320 corn. Plant Physiol 86: 359–363

362. Ogawa K, Tsuda M, Yamauuchi F, Yamaguchi I, Misato T (1976) Metabolism of 2-*sec*. butylphenol N-methyl carbamate (Bassa, BMPC) in rice plants and its degradation in soils. Pestic Sci 1: 219–224

363. Ohkawa H, Tsujii H, Ohkawa Y (1999) The use of cytochrome P450 genes to introduce herbicide tolerance in crops: a review. Pestic Sci 55: 867–874

364. O'Neill SD, Keith B, Rappaport L (1986) Transport of gibberellin A1 in cowpea membrane vesicles. Plant Physiol 80: 812–817

365. Opresko DM (1998) Toxicity summary for 2,4,6-trinitrotoluene. http://risk. lsd.ornl.gov/tox/ profiles/2_4_6_trinitrotoluene_f_V1.shtml.

366. Palazzo AJ, Leggett DC (1986) Effect and disposition of TNT in a terrestrial plant. J Environ Qual 15: 49–52

367. Paterson S, Mackay D, Tam D, Shiu WY (1990) Uptake of organic chemicals by plants: A review of processes, correlations and models. Chemosphere 21: 297–331

368. Paterson S, Mackay D, McFarlane C (1994) A model of organic chemical uptake by plants from soil and the atmosphere. Environ Sci Technol 28: 2259–2266

369. Pavlostathis SG, Comstock KG, Jacobson ME, Saunders FM (1998) Transformation of 2,4,6-trinitrotoluene by the aquatic plant *Myriophyllum spicatum*. Environ Toxicol Chem 17: 2266–2273

370. Pemadasa MA (1979) Stomatal responses to two herbicidal auxins. J Exp Bot 30: 267–276

371. Penner D, Early RW (1973) Effect of alachlor, butylate and chlorbromuron on carbofuran distribution and metabolism in barley and corn. Weed Sci 21: 360–366
372. Pereira JF, Splittstoesser WE, Hopen HJ (1971) Mechanism of intraspecific selectivity of cabbage to nitrofen. Weed Sci 19: 647–655
373. Perez-Maldonado IN, Diaz-Barriga F, de la Fuente H, Gonzalez-Amaro R, Calderon J, Yanez L (2004) DDT induces apoptosis in human mononuclear cells *in vitro* and is associated with increased apoptosis in exposed children. Environ Res 94: 38–46
374. Perkovich BS, Anderson TA, Kruger EL, Coats JR (1996) Enhanced mineralization of [^{14}C] atrazine in *Kochia scoparia* rhizospheric soil from a pesticide-contaminated site. Pestic Sci 46: 391–396
375. Pesticide chemictry and bioscience. The food–environment challenge. (1999) Brooks GT, Roberts TR (eds) The Royal Society of Chemistry, Cambridge
376. Pesticides literature review. Systematic review of pesticide human health effects. http://www.ocfp.on.ca/local/files/Communications/Current%20Issues/Pesticides/Final%20Paper%2023APR2004.pdf
377. Peterson FJ, Mason RP, Horspian J, Holtzman JL (1979). Oxygen-sensitive and insensitive nitroreduction by *Escherichia coli* and rat hepatic microcosomes. J Biol Chem 254: 4009–4014
378. Peterson JS, Cairns CB (2003) Toxicity, nitrous dioxide. http://www.emedicine. com/emerg/ topic847.htm
379. Peterson MM, Horst GL, Shea PJ, Comfort SD (1998) Germination and seedling development of switchgrass and smooth bromegrass exposed to 2,4,6-trinitrotoluene. Environ Pollut 99: 53–59
380. Peuke AD, Kopriva S, Rennenberg H (2004) Phytoremediation with the help of transgenic trees. In: Phytoremediation: environmental and molecular biological aspects. OECD workshop, Hungary, Abstr, p 33
381. Pidwirny M (1996) Formation of acid deposition. http://royal.okanagan.bc. ca/mpidwirn/ atmosphereandclimate/formacid dep.html
382. Pierrel MA, Batard Y, Kazmaier M, Mignotte-Vieux C, Durst F, Werck-Reichhart D (1994) Catalytic properties of the plant cytochrome P450 CYP73 expressed in yeast. Substrate specificity of a cinnamate hydroxylase. Eur J Biochem 224: 835–842
383. Pierzynski GM, Schnoor JL, Banks MK, Tracy JC, Licht LA, Erickson LE (1994) Vegetative remediation at superfund sites. Mining and its environment impact. Royal Soc Chem Issues in Environ Sci Technol. vol 1, pp 49–69
384. Pillai CG, Parthasarathy Weete JD, Davis DE (1977) Metabolism of atrazine by *Spartina alterniflora*. 1. Chloroform-soluble metabolites J Agric Food Chem 25: 852–856
385. Pivetz B, Cochran R, Huling S (1997) Phytoremediation of PCP and PAH-contaminated soil. Poster 54. In: 12th Annu Conf Hazardous Waste Res. Kansas City. Abstr, http://www.engg.ksu.edu/HSRC/97abstracts/p54.html
386. Prasad S, Ellis BE (1978) *In vivo* characterization of catechol ring-cleavage in cell cultures of *Glycine max*. Phytochemistry 17: 187–193

387. Preuss A, Rieger PG (1995) Anaerobic transformation of 2,4,6-TNT and other nitroaromatic compounds. In: Spain JC (ed) Biodegradation of nitroaromatic compounds. Plenum Press, New York, pp 69–85
388. Pridham JB (1958) Metabolism of phenolic compounds by the broad bean *Vicia faba*. Nature 182: 795–800
389. Pruidze GK, Mchedlishvili NI, Omiadze NT, Gulua LK, Pruidze NG (2003) Multiple forms of phenol oxidase from Kolkhida tea leaves (*Camellia sinensis* L.) and *Mycelia sterilia* IBR 35219/2 and their role in tea production. Food Res Int 36: 587–595
390. Psaras GK, Christodoulakis NS (1987) Air pollution effects on the structure of *Citrus aurantium* leaves. Bull Environ Contam Toxicol 39: 474–480
391. Qiu X, Shah SI Kendall EW, Sorensen DL, Sims RC, Engelke MC (1994) Grass-Enhanced bioremediation for clay soils contaminated with polynuclear aromatic hydrocarbons. In: Anderson TA, Coats JR (eds) Bioremediation through rhizosphere technology. ACS Symp Ser. Am Chem Soc, Washington, vol 563, pp 142–157
392. Qiu X, Leland TW, Shah SI, Sorensen DL, Kendall EW (1997) Field study: grass remediation for clay soil contaminated with polycyclic aromatic hydrocarbons. In: Kruger EL, Anderson TA, Coats JR (eds) Phytoremediation of soil and water contaminants. Am Chem Soc, Washington, pp 186–199
393. Rahman A, Matthews LJ (1979) Effects of soil organic matter on the phytotoxicity of thirteen S-triazine herbicides. Weed Sci 27: 158–167
394. Ramina A, Pimpini F, Boniolo A, Bergamasco F (1979) (8-^{14}C) Benzylaminopurine translocation in *Phaseolus vulgaris*. Plant Physiol 63: 294–298
395. Ramsden JJ (1993) Partial molar volume of solutes in bilayer lipid membranes. J Phys Chem 97: 4479–4483
396. Ramsden JJ (1993) Partition coefficients of drugs in bilayer lipid membranes. Experientia 49: 688–692
397. Raskin IP, Nanda Kumar BA, Dushenkov S, Blaylock MJ, Salt D (1994) Phytoremediation – using plants to clean up soils and waters contaminated with toxic metals. Emerging Technologies in Hazardous Waste Management VI, ACS Industrial & Engineering Chemistry Division Special Symp. vol 1, Atlanta
398. Reddy KN, Bendixen LE (1989) Toxicity, absorption, and translocation of soil-applied chlorimuron in yellow and purple nutsedge *(Cyperus esculentus and C. rotundus)*. Weed Sci 37: 147–151
399. Reeves RD, Brooks RR (1983) Hyperaccumulation of lead and zinc by two metallophytes from mining areas of Central Europe. Environ Pollut Ser A 31: 277–285
400. Reiger PG, Knackmuss HJ (1995) Basic knowledge and perspectives on biodegradation of 2,4,6-trinitrotoluene and related nitroaromatic compounds in contaminated soil. In: Spain JC (ed) Biodegradation of nitroaromatic compounds. Plenum Press, New York, pp 1–20
401. Reilley KA, Banks MK, Schwab AP (1996) Dissipation of polycyclic aromatic hydrocarbons in the rhizosphere. J Environ Qual 25: 212–219

402. Rhodes R (1977) Metabolism of (2-^{14}C)terbacil in alfalfa. J Agric Food Chem 25:1066–1070
403. Rice PJ, Anderson TA, Coats JR (1997) Phytoremediation of herbicide-contaminated surface water with aquatic plants. In: Kruger EL, Anderson TA, Coats JR (eds) Phytoremediation of soil and water contaminants. Am Chem Soc, Washington, pp 133–151
404. Rieder G, Buchholtz KP, Kust CA (1970) Uptake of herbicides by soybean seed. Weed Sci 18: 101–120
405. Rigby NM, McDougall AJ, Needs PW, Selvendran RR (1994) Phloem translocation of a reduced oligogalacturonide in *Ricinus communis* L. Planta 193: 536–541
406. Rivera R, Medina VF, Larson SL, McCutcheon SC (1998) Phytotreatment of TNT-contaminated groundwater. J Soil Contam 7: 511–529
407. Roberts JR (1999) Metal toxicity in children. In: Training manual on pediatric environmental health: putting it into practice. Children's Environmental Health Network. Emeryville, http://www.cehn.org/cehn/trainingmanual/pdf/manual-full.pdf
408. Robidoux PY, Hawari J, Thiboutot S, Ampleman G, Sunahara GI (1999) Acute toxicity of 2,4,6-trinitrotoluene in earthworm (*Eisenia andrei*). Ecotoxicol Environ Saf 44: 311–321
409. Robineau T, Batard Y, Nedelkina S, Cabello-Hurtado F, LeRet M, Sorokine O, Didierjean L, Werck-Reichhart D (1998) The chemically inducible plant cytochrome P450 CYP76B1 actively metabolizes phenylureas and other xenobiotics. Plant Physiol 118: 1049–1056
410. Robocker WC, Zamora BA (1976) Translocation and metabolism of dicamba in western bracken. Weed Sci 24: 435–441
411. Rodríguez-López JN, Tudela J, Varón R, Fenoll LG, García-Carmona F, García-Cánovas F (1992) Analysis of a kinetic model for melanin biosynthesis pathway. J Biol Chem 267: 3801–3810
412. Rodríguez-López JN, Fenoll LG, García-Ruiz PA, Varón R, Tudela J, Thorneley RNF, García-Cánovas F (2000) Stopped-flow and steady-state study of the diphenolase activity of mushroom tyrosinase. Biochemistry 39: 10497–10506
413. Rodríguez-López JN, Lowe DJ, Hernández-Ruiz J, Hiner ANP, García-Cánovas F, Thorneley RNF (2001) Mechanism of reaction of hydrogen peroxide with horseradish peroxidase: identification of intermediates in the catalytic cycle. J Am Chem Soc 123: 11838–11847
414. Rolston MP, Robertson AG (1975) Some aspects of absorption, translocation, and metabolism of ^{14}C-picloram in gorse. Proc NZ Weed Pestic Control Conf 28: 54–59
415. Roy S, Hanninen O (1994) Pentachloroohenol: uptake/elimination kinetics and metabolism in aquatic plant *Eichornia crassipes*. Environ Toxicol Chem 13: 763–773
416. Rubin B, Eshel Y (1977) Absorption and translocation of terbutryn and fluometuron in cotton (*Gossypium hirsutum*) and snapbean (*Phaseolus vulgaris*) Weed Sci 25: 499–505

417. Rudolf W (1994) Concentration of air pollutants inside cars driving on highways and in downtown areas. Sci Total Environ 146/147: 433–444
418. Rusin VY (1988) Lead and its compounds. In: Filov VA (Ed) Harmful chemical substances (in Russian). Khimiya, Leningrad, pp 415–436
419. Sachse B, Lötzsch K, Schulze EF (1974) Untersuchungen zur Aufnahme and Translokation des Fungizides Pyracarbolid in Kaffee. Meded Fac landbauwwetensch. Rijksuniv Gent 39: 1145–1153
420. Salaün JP (1991) Metabolization de xenobiotiques par des monooxygenases a cytochrome P-450 chez les plantes. Oceanis 17: 459–474
421. Salaün JP, Helvig C (1995) Cytochrome P450-dependent oxidation of fatty acids. In: Durts F, O'Keef DP (eds) Drug metabolism and drug interactions. Freund Publishing House, England, pp 12–49
422. Salt DE, Blaylock M, Nanda Kumar PBA, Dushenkov VP, Ensley BD, Chet I, Raskin I (1995) Phytoremediation: a novel strategy for the environment using plants. Biotechnology 13: 468–474
423. Salt DE, Smoth RD, Raskin I (1998) Phytoremediation. Annu Rev Plant Physiol Mol Biol 49: 643–668
424. Samoiloff M (1998) Benzene toxicity. Benzene and derivatives. Organic chemistry 3. http://xnet.rrc.mb.ca/martins/Organic%203/benzene.htm
425. Sánches-Ferrer A, Rodrígez-López JN, García-Cánovas F, García-Carmona F (1994) Tyrosinase: a comprehensive review of its mechanism. Biochim Biophys Acta 1247: 1–11
426. Sandermann H (1987) Pestizid-Rückstände in Nahrungspflanzen. Die Rolle des pflanzlichen Metabolismus. Naturwissenschaften 74: 573–578
427. Sandermann H (1988) Mutagenic activation of xenobiotics by plant enzymes. Mutation Res 197: 183–194
428. Sandermann H (1994) Higher plant metabolism of xenobiotics: the "green liver" concept. Pharmacogenetics 4: 225–241
429. Sandermann H, Scheel D, Trenck T (1983) Metabolism of environmental chemicals by plants. Copolymerization into lignin. J Appl Polymer Sci 37: 407–420
430. Sargent JA, Blackman GE (1972) Studies on foliar penetration. 9. Patterns of penetration of 2,4-dichlorophenoxyacetic acid into the leaves of different species. J Exp Bot 23: 830–839
431. Saxe H (1996) Physiological and biochemical tools in diagnosis of forest decline and air pollution injury to plants. In: Yunus M, Iqbal M (eds) Plant responses to air pollution. Wiley, New York Chichester
432. Schenckman JB, Cinti DL, Moldeus PW (1973) The mitochondrial role in hepatic cell mixed-function oxidations. Ann NY Acad Sci 212: 420–427
433. Scheunert I, Klein W (1985) Predicting the movement of chemicals between environments (air-water-soil-biota). In: Sheehan P, Korte F, Klein W, Bourdeau P (eds) Appraisal of tests to predict the environment behaviour of chemicals. Wiley, New York Chichester, pp 285–332
434. Schlee D, Thoringer C, Tintemann H (1994) Purification and properties of glutamate-dehydrogenase in Scots pine (*Pinus sylvestris*) needles. Physiol Plant 92: 467–472

435. Schmitt R, Kaul J, Trenck T, Schaller E, Sandermann H (1985) β-D-Glucosyl and O-malonyl-β-D-glucosyl conjugates of pentachlorophenol in soybean and wheat: Identification and enzymatic synthesis. Pestic Biochem Physiol 24: 77–85

436. Schnabel WE, Dietz AC, Burken JG, Schnoor JL, Alvarez PJ (1997) Uptake and transformation of trichloroethylene by edible garden plants. Water Res 31: 816–824

437. Schnoor JL, Dee PE (1997) Phytoremediation. Technology Evaluation Report TE-98-01. Ground-Water Remediation Technologies Analysis Center. Ser E. Iowa City

438. Schnoor JL, Licht LA, McCutcheon SC, Wolfe NL, Carreira LH (1995) Phytoremediation of organic and nutrient contaminants. Environ Sci Technol 29: 318A–323A

439. Schoenmuth BW, Pestemer W (2004) Dendroremediation of trinitrotoluene (TNT) Part 1: Literature overview and research concept. Environ Sci Pollut Res 11: 273–278

440. Schoenmuth BW, Pestemer W (2004) Dendroremediation of trinitrotoluene (TNT) Part 2: Fate of radio-labelled TNT in trees. Environ Sci Pollut Res 11: 331–339

441. Schönherr J, Bukovac MJ (1972) Penetration of stomata by liquids. Dependence on surface tension, wettability, and stomatal morphology. Plant Physiol 49: 813–823

442. Schönherr J, Bukovac MJ (1978) Foliar penetration of succinic acid-2,2-dimethylhydrazide: mechanism and rate limiting step. Physiol Plant 42: 243–249

443. Schuler MA (1996) Plant cytochrome P450 monooxygenases. Crit Rev Plant Sci 15: 235–284

444. Schultz ME, Burnside OC (1980) Absorption, translocation and metabolism of 2,4-D and glyphosate in hemp dogbane (*Apocynum cannabinum*). Weed Sci 28: 13–19

445. Schuphan I, Ebmz W (1978) Metabolism and balance studies of (^{14}C) monolinuron after use in spinach followed by cress and potato cultures. Pestic Biochem Physiol 9: 107–113

446. Schwab AP, Al Assi AA, Banks MK (1998) Adsorption of naphthalene onto plant roots. J Environ Qual 27: 220–224

447. Scott HD, Phillips RE (1971) Diffusion of herbicides to seed. Weed Sci 19: 128–137

448. Scott HD, Phillips RE (1973) Absorption of herbicides by soybean seed. Weed Sci 21: 71–79

449. Seidel K, Kickuth R (1967) Exkretion von Phenol in der Phylosphäre von *Scirpus lacustris* L. Naturwissenschaften 52: 517–525

450. Selifonov SA, Chapman PJ, Akkerman SB, Gurst JE, Bortiatynski JM, Nanny MA, Hatcher PG (1998) Use of ^{13}C nuclear magnetic resonance to assess fossil fuel degradation: Fate of [1-^{13}C] acenaphthene in creosote polycylic aromatic compound mixtures degraded by bacteria. Appl Environ Microbiol 64: 1447–1453

451. Sens C, Sheidemann P, Werner D (1999) The distribution of ^{14}C-TNT in different biochemical compartments of the monocotyledoneous *Triticum aestivum*. Environ Pollut 104: 113–119

452. Shang TQ, Newman LA, Gordon MP (2003) Fate of trichloroethylene in errestrial plants. In: McCutcheon SC, Schnoor JL (eds) Phytoremediation. Transformation and control of contaminats. Wiley-Interscience, Hoboken, New Jersey, pp 529–560

453. Sharma MP, Vanden Born WH (1970) Foliar penetration of picloram and 2,4-D in aspen and balsam poplar. Weed Sci 18: 57–65

454. Sharma D, Bhanlvaj R, Maheshwari V (1989) Inhibition of photosynthesis by oxyfluorfen. Curr Sci 58: 1334–1336

455. Shida T, Homma Y, Misaio T (1975) Absorption, translocation and degradation of N-lauoryl-L-valine in plants. 6. Studies on the control of plant diseases by amino acid derivatives. J Agric Chem Soc Jap 49: 409–418

456. Shimabukuro RH, Hoffer BL (1991) Metabolism of diclofop-methyl in sensitive and resistant biotypes of *Lolium rigidum*. Pestic Biochem Physiol 39: 251–260

457. Shimabukuro RH, Walsh WC, Jacobson A (1987) Aryl-O-glucoside of diclofop: a detoxification product in wheat shoots and wild oat suspension diclofop. J Agric Food Chem 35: 393–399

458. Shimabukuro RH, Walsh WC, Hoffer BL (1989) The absorption, translocation and metabolism of difenopenten-ethyl in soybean and wheat. Pestic Biochem Physiol 33: 57–68

459. Shinohara A, Kamataki T, Ichimura Y, Opochi H, Okuda K, Kato R (1984) Drug oxidation activities of horse-redish peroxidase, myoglobin and cytochrome P-450$_{cam}$ reconstituted with synthetic hemes. Jap J Pharmacol 45: 107–114

460. Shiu WY, Mackay D (1997) Henry's law constants of selected aromatic hydrocarbons, alcohols, and ketones. J Chem Eng Data 42: 27–30

461. Shutte R, Goffmann GP (1975) Metabolism of herbicides: derivatives of diphenyl ether in oat. In: Mechanism of plant herbicide and synthetic growth regulators actions and their further fate in biosphere (in Russian). Materials of XI Symp. Pushchino, pp 129–133

462. Sicbaldi F, Sacchi GA, Trevisan M, Del Re AAM (1997) Root uptake and xylem translocation of pesticides from different chemical classes. Pestic Sci 50: 111–119

463. Siciliano SD, Germida JJ (1997) Bacterial inoculants of forage grasses enhance degradation of 2-chlorobenzoic acid in soil. Environ Toxicol Chem 16: 1098–1104

464. Siciliano SD, Roy R, Greer CW (2000) Reduction in denitrification activity in field soils exposed to long term contamination by 2,4,6-trinitrotoluene (TNT). FEMS Microbiol Ecol 32: 61–68

465. Siegel BZ (1993) Plant peroxidases – an organismic perspective. J Plant Growth Regul 12: 303–312

466. Simon EW, Beavers H (1954) The effect of pH on the biological activities of weak acids and bases. 1. The most usual relationship between pH and activity. New Phytol 51: 163–169
467. Sincero APP, Sincero GA (1999) Environmental engineering: a design approach. Prentice-Hall, New Jersy
468. Smeltzer SC, Bare BG (1996) Brunner and Suddharth's textbook of medical-surgical nursing, 8th edn. Lippincott-Raven Publishers, Pittsburg
469. Smith AE, Phaull SC, Emmatty DA (1989) Metribuzin metabolism by tomato cultivars with low, medium and high levels of tolerance to metribuzin. Pestic Biochem Physiol 35: 284–290
470. Smith G, Neyra C, Brennan E (1990) The relationship between foliar injury, nitrogen metabolism and growth parameters in ozonated soybeans. Environ Pollut 63: 79–83
471. Söchtig H (1964) Beeinflussung des Stoffwechsels der Pflanzen durch Humus und seine Bestandteile und die Auswirkung auf Wachstum und Ertrag. Landbauforsch Völkenrode 14: 9–15
472. Song WY, Sohn EJ, Martinoia E, Lee YJ, Yang Y-Y, Jasinski M, Forestier C, Hwang I, Lee Y (2003) Engineering tolerance and accumulation of lead and cadmium in transgenic plants. Nat Biotechnol 21: 914–919
473. Spain JC, Hughes JB, Knackmuss H-J (eds) (2000) Biodegradation of nitroaromatic compounds and explosives. Lewis Publishers, Boca Raton London New York Washington
474. Spence RD, Rykiel EJ, Shrape PJH (1990) Ozone alters carbon allocation in loblolly pine: assessment with carbon-11 labeling. Environ Pollut 64: 93–106
475. Spencer CI, Yuill KH, Borg JJ, Hancox JC, Kozlowski RZ (2001) Actions of pyrethroid insecticides on sodium currents, action potentials, and contractile rhythm in isolated mammalian ventricular myocytes and perfused hearts. Pharmacol 298: 1067–1082
476. Spencer WF Farmer WJ, Cliath MM (1973) Pesticide volatilization. J Residue Rev 49: 1–47
477. Spencer WF, Cliath MM, Jury WA, Zhang L-Z (1988) Volatilization of organic chemicals from soil as related to Henry's law constants. J Environ Qual 17: 504–509
478. Stahl JD, Aust SD (1995) Biodegradation of 2,4,6-trinitrotoluene by the white rot fungus *Phanerochaete chrysosporium*. In: Spain JC (ed) Biodegradation of nitroaromatic compounds. Plenum Press, New York, pp 117–134
479. Steinert WG, Stnizke JF (1977) Uptake and phytotoxicity of tebuthiuron. Weed Sci 25: 390–396
480. Sterling TM, Blake NE (1988) Use of soybean *(Glycine max)* and velvetleaf *(Abutilon theophrasti)* suspension-cultured cells to study bentazon metabolism. Weed Sci 36: 558–565
481. Sterling TM, Blake NE (1989) Differential bentazon metabolism and retention of bentazon metabolites by plant cell cultures. Pestic Biochem Physiol 34: 39–48

482. Sterling TM, Blake NE (1990) Bentazon uptake and metabolism by cultured plant cells in the presence of monooxygenase inhibitors and cinnamic acid. Pestic Biochem Physiol 38: 66–75

483. Stiborova M, Anzenbacher P (1991) What are the principal enzymes oxidizing the xenobiotics in plants: cytochrome P-450 or peroxidase? Gen Physiol 10: 209–216

484. Still GG, Mansager ER (1973) Metabolism of isopropyl-3-chlorocarbanilate by cucumber plants. J Agric Food Chem 21: 697–700

485. Stoker SH, Seager SL (1982) Pollution by organic compounds (oil, pesticides, surfactants) (in Russian) In: Bockris JOM (Ed) Environmental chemistry. Khimiya, Moskow, pp 346–370

486. Sugumaran M, Duggaraju R, Generozova F, Ito S (1999) Insect melanogenesis. II. Inability of *Manduca* phenoloxidase to act on 5,6-dihydroxyindole-2-carboxylic acid. Pigm Cell Res 12: 118–125

487. Sun GF, Pi JB, Li B, Guo XY, Yamavchi H, Yoshida T (2000) Introduction of present arsenic research in China. Paper presented at 4th Int Conf on Arsenic Exposure and Health Effects. Soc Geochem and Health, San Diego

488. Susarla S, Medina VF, McCutcheol SC (2002) Phytoremediation: an ecological solution of organic chemical contamination. Ecolog Eng 18: 647–658

489. Sweetser PB, Schow GS, Hutchinson JM (1982) Metabolism of chlorsulfuron by plants: biological basis for selectivity of a new herbicide for cereals. Pestic Biochem Physiol 17: 18–23

490. Tabata M, Ikeda F, Haraoka N, Konoshima M (1976) Glucosylation of phenolic compounds by *Datura innoxia* suspension cultures. Phytochemistry 15: 1225–1229

491. Talekar NS, Lee EM, Sun LT (1977) Absorption and translocation of soil and foliar applied ^{14}C-carbofuran and ^{14}C-phorate in soybean and mung bean seeds. J Econ Entomol 70: 685–688

492. Tanaka FS, Hoffer BL, Simabukuro RH (1990) Identification of the isomeric hydroxylated metabolites of methyl-2-[4-(2,4-dichlorophenoxy)phenoxy] propanoate (diclofop-methyl) in wheat. J Agric Food Chem 38: 599–603

493. Tateoka TN (1970) Studies on the catabolic pathway of protocatechic acid in mung bean seedlings. Bot Mag (Tokyo) 83: 49–54

494. Taylor H, Wain R (1978) Studies of plant growth-regulating substances. 52. Growth retardation by 3,5-dichlorophenoxyethylamine and 3,5-dichloro-phenoxybutyric acid arising from their conversion to 3,5- dichlorophenoxyacetic acid in tomato plants. Ann Appl Biol 89: 271–277

495. Taylorson RB (1979) Response of weed seeds to ethylene and related hydrocarbons. Weed Sci 27: 27–36

496. Tenga AZ, Ormrod DP (1990) Diminished greenness of tomato leaves exposed to ozone and post-exposure recovery of greenness. Environ Pollut 64: 29–41

497. Thompson PL, Ramer LA, Schnoor JL (1998) Uptake and transformation of TNT by hybrid poplar trees. Environ Sci Technol 32: 975–980

498. Tkhelidze P (1969) Oxidative transformation of benzene and toluene in vine grapes (in Russian). Bull Georg Acad Sci 56: 697–700

499. Tolls J, de Graaf I, Thijssen MATC, Haller M Sijm DTHM (1997) Bioconcentration of LAS: Experimental determination and extrapolation to environmental mixtures. Environ Sci Technol 31: 3426–3431
500. Topp E, Scheunert I, Korte F (1989) Kinetics of the uptake of ^{14}C-labelled chlorinated benzene from soil by plants. Ecotoxicol Environ Safety 17: 157–166
501. Torres LG, Santacruz G Bandala ER (1999) Biodegradation of 2,4-D and DDT in high concentrations in low-cost packaging biofilters. In: Alleman BC, Leeson A (eds) Bioremediation of nitroaromatic and haloaromatic compounds. Battelle Press, Columbus
502. Trenk T, Sandermann H (1978) Metabolism of benzo[a]pyrene in cell suspension cultures of parsley (*Petroselinum hortense,* Hoffm.) and soybean (*Glycine max* L.). Planta 141: 245–251
503. Trenck T, Sandermann H (1980) Oxygenation of benzo[a]pyrene by plant microsomal fractions. FEBS Lett 119: 227–231
504. Trust BA, Muller JG, Goffin RB, Gifuentes LA (1995) Biodegradation of fluoranthrene as monitored using stable carbon isotopes. In: Hinchee RE, Douglas GS, Ong SK (eds) Monitoring and verification of bioremediation. Battelle Press, Columbus, pp 223–239
505. Tsao DT (2003) Phytoremediation. Advances in biochemical engineering and biotechnology. Springer, Berlin Heidelberg New York
506. Turcsanyi G (1992) Plant cells and tissues as indicators of environmental pollution. In: Kovács M (ed) Biological indicators in environmental protection. Ellis Horwood, New York
507. Tyree MT, Peterson CA, Edgington LV (1979) A simple theory regarding ambimobility of xenobiotics with special reference to the nematicide oxamyl. Plant Physiol 63: 367–374
508. Ugrekhelidze D (1976) Metabolism of exogenous alkanes and aromatic hydrocarbons in plants (in Russian). Metsnieraba, Tbilisi
509. Ugrekhelidze D, Arziani B (1980) Peptide conjugates of α-naphthol and o-nitrophenol in plants (in Russian). Bull Georg Acad Sci 100: 686–689
510. Ugrekhelidze D, Durmishidze S (1980) The biosphere chemical pollution and plant (in Georgian). Metsniereba, Tbilisi
511. Ugrekhelidze D, Durmishidze S (1984) Penetration and detoxification of organic xenobiotics in plants (in Russian). Metsniereba, Tbilisi
512. Ugrekhelidze D, Kavtaradze L (1970) The question of metabolism of α-naphthol in higher plants (in Russian). Bull Georg Acad Sci 57: 465–469
513. Ugrekhelidze D, Arziani B, Mithaishvili T (1983) Peptide conjugates of exogenous monoatomic phenols in plants (in Russian). Fiziol Rast (Moscow) 30: 102–107
514. Ugrekhelidze D, Phiriashvili V, Mithaishvili T (1986) Uptake of salicylic acid and aniline by pea roots (in Russian). Fiziol Rast (Moscow) 33: 165–170
515. Ugrekhelidze D, Korte F, Kvesitadze G (1997) Uptake and transformation of benzene and toluene by plant leaves. Ecotoxicol Environ Saf 37: 24–28

516. Van de Pas BA, Smidt H, Hagen WR, van der Oost J, Schraa G, Stams AJM, de Vos WM (1999) Purification and molecular characterization of ortho-chlorophenol reductive dehalogenase, a key enzyme of halorespiration in *Desulfitobacterium dehalogenans*. J Biol Chem 274: 20287–20292

517. Vanderford M, Shanks JV, Hughes JB (1997) Phytotransformation of trini-trotoluene (TNT) and distribution of metabolic products in *Myriophyllum aquaticum*. Biotechnol Lett 3: 277–280

518. Veeranjaneyulu K, Charlebois D, N'soukpoe-Kossi CN, Leblanc RM (1990) Effect of sulfur dioxide and sulfite on photochemical energy storage of iso-lated chloroplasts – a photoacoustic study. Environ Pollut 65: 127–139

519. Vias SC, Sinha OK, Josh LK (1976) Systemic uptake and translocation of bavistin and calixin in peanut *(Arachis hypogaea* L.). Pesticides 10: 32–38

520. Wagner SC, Zablotowicz RM (1997) Utilization of plant material for reme-diation of herbicide-contaminated soils. In: Kruger EL, Anderson TA, Coats JR (eds) Phytoremediation of soil and water contaminants. Am Chem Soc, Washington, pp 65–76

521. Walter H, Breckle SW (1991) Ökologie der Erde 1 – Ökologische Grundla-gen in globaler Sicht, 2. Aufl. Fischer, Stuttgart

522. Weber JB, Weed SB, Waldrep TW (1974) Effect of soil constituents on her-bicide activity in modified soil field plots. Weed Sci 22: 454–463

523. Weimer MR, Swisher BA, Vogel KP (1988) Metabolism as a basis for dif-ferential atrazine tolerance in warm-season forage grasses. Weed Sci 36: 436–440

524. Wellburn FAM, Lau KK, Milling PMK, Wellburn AR (1996) Drought and air-pollution affect nitrogen cycling and free-radical scavenging in *Pinus halepensis* (Mill). J Exp Bot 47: 1361–1367

525. Wetzel A, Sandermann H (1994) Plant biochemistry of xenobiotics: isolation and characterization of a soybean O-glucosyltransferase of DDT metabolism. Arch Biochem Biophys 314: 323–328

526. Whelton BD, Peterson DP, Moretti ES, Dare H, Bhattacharyya MH (1997) Skeletal changes in multiparous, nulliparous and ovariectomized mice fed ei-ther a nutrient-sufficient or -deficient diet containing cadmium. Toxicology 119: 103–121

527. Whetten R, Sederoff R (1995) Lignin biosynthesis. Plant Cell 7: 1001–1013

528. White RH, Liebel RA, Hymowitz T (1990) Examination of 2,4-D tolerance in perennial *Glycine* species. Pestic Biochem Physiol 38: 153–161

529. Wichman JR, Byrnes WR (1975) Uptake, distribution and degradation of si-mazine by black walnut and yellow poplar seedlings. Weed Sci 23: 448–454

530. Wilcut JM, Wehtje GR, Patterson MG, Cole TA, Hicks TV (1989) Absorp-tion, transformation and metabolism of foliar-applied chlorimuron in soy-bean *(Glycine max),* peanut *(Arachis hypogaea)* and selected weeds. Weed Sci 37: 175–180

531. Wilkner R, Sandermann H (1989) Plant metabolism of chlorinated anilines: isolation and identification of N-glucosyl and N-malonyl conjugates. Pestic Biochem Physiol 33: 239–248

532. Williams J, Miles R, Fosbrook C, Deardorff T, Wallace M West B (2000) Phytoremediation of aldrin and dieldrin: a pilot-scale project. In: Wickramanayake GB, Gavaskar AR, Gibbs JT Means JL (eds) Case studies in the remediation of chlorinated and recalcitrant compounds. Battelle Press, Columbus

533. Wills GDH, Scriven FM (1979) Metabolism of geraniol by apples in relation to the development of storage breakdown. Phytochemistry 18: 785–790

534. Wilson L, Williamson T, Gronowski J, Gentile GI, Gentile JM (1994) Characterization of 4-nitro-o-phenylenediamine activities by plant systems. Mutation Res 307: 185–193

535. Wiltse CC, Rooney WL, Chen Z, Schwab AP, Banks MK (1998) Greenhouse evaluation of agronomic and crude oil-phytoremediation potential among alfalfa genotypes. J Environ Qual 27: 169–173

536. Winner WE (1994) Mechanistic analysis of plant-responses to air-pollution. Ecol Applic 4: 651–661

537. Wolfe NL, Hoehamer CF (2003) Enzymes used by plants and microorganisms to detoxify organic compounds. In: McCutcheon SC, Schnoor JL (eds) Phytoremediation. Transformation and control of contaminants. Wiley-Interscience, Hoboken, New Jersey, pp 159–188

538. Yalpani N, Silverman P, Wilson MA, Kleier DA, Raskin I (1991) Salicylic acid is a system signal and an inducer of pathogenesis-related proteins in virus-infected tobacco. Plant Cell 3: 809–818

539. Yanez L, Borja-Aburto VH, Rojas E, de la Fuente H, Gonzalez-Amaro R, Gomez H, Jongitud AA, Diaz-Barriga F (2004) DDT induces DNA damage in blood cells. Studies in vitro and in women chronically exposed to this insecticide. Environ Res 94: 18–24

540. Yin SN, Li GL, Tain FD (1987) Leukaemia in benzene workers: a retrospective cohort study. Br J Ind Med 44: 124–128

541. Yoon JM, Oh BT, Just CL, Schnoor JL (2002) Uptake and leaching of octahydro-1,3,5,7-tetranitro-1,3,5,7-tetrazocine by hybrid poplar trees. Environ Sci Technol 36: 4649–4655

542. Young P (1996) The new science of wetland restoration. Environ Sci Technol 30: 292A–296A

543. Yu MH, Sakurai S (1995) Diisopropylfluorylphosphate (DFP)-hydrolyzing enzymes in mung bean (Vigna radiata) seedlings. Environ Sci (Tokyo) 3: 103–111

544. Zaalishvili G, Khatisashvili G, Ugrekhelidze D, Gordeziani M, Kvesitadze G (2000) Plant potential for detoxification (Review). Appl Biochem Microbiol 36: 443–451

545. Zaalishvili G, Lomidze E, Buadze O, Sadunishvili T, Tkhelidze P, Kvesitadze G (2000) Electron microscopic investigation of benzidine effect on maize root tip cell ultrastructure, DNA synthesis and calcium homeostasis. Int Biodeterior Biodegrad 46: 133–140

546. Zaalishvili G, Sadunishvili T, Scala R, Laurent F, Kvesitadze G (2002) Electron microscopic investigation of nitrobenzene distribution and effect on plant root tip cells ultrastructure. Ecotoxicol Environ Saf 52: 190–197

547. Zenno S, Kobori T, Tanokura M, Saigo K (1998) Conversion of *NfsA*, the major *Escherichia coli* nitroreductase, to a flavin reductase with an activity similar to that of Frp, a flavin reductase in *Vibrio harveyi*, by a single amino acid substitution. J Bacteriol 180: 422–425

548. Zhu Y-P, Rosen MJ, Morall SW, Tolls J (1998) Surface properties of linear alkylbenzene sulfonates in hard river water. J Surfactants and Detergents 1: 187–193

549. Zimmerlin A, Durst F (1992) Aryl hydroxylation of the herbicide diclofop by a wheat cytochrome P-450 monooxygenase. Plant Physiol 100: 868–873

Index